Postdisciplinary Studies in Discourse

Series Editor
Johannes Angermuller
Centre for Applied Linguistics
University of Warwick
Coventry, UK

Postdisciplinary Studies in Discourse engages in the exchange between discourse theory and analysis while putting emphasis on the intellectual challenges in discourse research. Moving beyond disciplinary divisions in today's social sciences, the contributions deal with critical issues at the intersections between language and society.

Edited by Johannes Angermuller together with members of DiscourseNet, the series welcomes high-quality manuscripts in discourse research from all disciplinary and geographical backgrounds. DiscourseNet is an international and interdisciplinary network of researchers which is open to discourse analysts and theorists from all backgrounds.

More information about this series at
http://www.palgrave.com/gp/series/14534

M. Cristina Caimotto

Discourses of Cycling, Road Users and Sustainability

An Ecolinguistic Investigation

palgrave
macmillan

M. Cristina Caimotto
Culture, Politics and Society
University of Turin
Turin, Italy

Postdisciplinary Studies in Discourse
ISBN 978-3-030-44025-1 ISBN 978-3-030-44026-8 (eBook)
https://doi.org/10.1007/978-3-030-44026-8

Cover illustration: Jani Riekkinen/Alamy Stock Photo

This Palgrave Macmillan imprint is published by the registered company Springer Nature Switzerland AG
The registered company address is: Gewerbestrasse 11, 6330 Cham, Switzerland

For Vanessa, whose smile when she rides her balance bicycle says it all.

Preface

This book is the result of an experiment, the attempt we are sometimes drawn to, of bringing together what seem to be very different aspects of one's own life. We thus throw a new light and a new perspective on what had appeared very familiar until the day before the idea dawned in our mind. This book is conceived as part of the Critical Discourse Studies approach. One of its tenets is that the researcher's motivation lies not in an interest in language for its own sake but an interest in how a social or political issue is played out in language. As van Dijk (2001, 96) puts it 'Critical Discourse Analysis is biased—and proud of it', as—unlike other scholarship—the researcher's own sociopolitical position is not denied but explicitly defined and defended. My advocacy activity brought me to form the plan for a linguistic investigation of the discourse of everyday cycling that, thus far, has not been analysed from a CDS perspective. The city where I live—Turin, Italy—has a history of car manufacturing and was designed to accommodate for people moving around mainly in their private car. But since 2010, a group of local citizens gave life to the largest parade[1] in Italy to advocate for better conditions for cyclists, promote the advantages of everyday cycling and ask the administration to foster change towards a more sustainable modal mix. I have always been

[1]Bike Pride FIAB Torino, a local advocacy association federate of the national FIAB (Italian Federation Environment and Bicycle) and the ECF (European Cyclists' Federation) was born in 2013 to reunite the citizens who had been organising the 'Bike Pride' parade since 2010.

passionate about slow, recreational cycling—a passion one can hardly share in a city where the car is dominant in every possible sense and in a country with an important sport cycling history, where most people conceive cycling as a demanding sport activity. Having used a bicycle as my main means of transport through most of my adult life, I was naturally drawn towards the advocacy group and I soon found myself acting as president of the advocacy association during the political campaign for the mayor's election in 2016.

Being a linguist, I observed the discourses deployed to either promote or reject an increase in bicycle use from a language perspective and I frequently had a feeling that the issues being debated reached far beyond the practicalities being discussed. This book aims to transform that feeling into an analysis grounded in linguistics. Drawing from a cross-disciplinary approach, disentangling, pinning down and observing the various discourses that are drawn upon when cycling is discussed—liberty, public property, social class, norm and deviancy, safety, stigma, discrimination, hate and human rights—to name but a few. The texts under scrutiny are in English, because English linguistics is my field of research and because this allows me to step back and analyse discourses that I am familiar with—due to my advocacy activity—but with which I am not directly involved. This would not be true if I were analysing the texts produced by the local media, policymakers or—even closer—local advocacy groups.

Given its experimental nature, the aim of this work is to offer new insights, to look at the issue from a new and refreshing perspective. It does not aim to offer definitive solutions, but to draw attention to the topic, presenting observations that can be the starting point of an interdisciplinary debate around what can be done discursively to improve the effectiveness of any policy promoting cycling and perhaps environment-friendly policies more in general. The analysis offers advocacy groups and policymakers increased awareness of discourses that are not evident on the surface and offers linguists a new object of study, full of complexity and fascination.

Turin, Italy M. Cristina Caimotto

Reference

van Dijk, T. A. (2001). Multidisciplinary CDA: A Plea for Diversity. In R. Wodak & M. Meyer (Eds.), *Methods of Critical Discourse Analysis* (pp. 95–120). London: Sage.

Acknowledgments

I am grateful to my colleagues Michelangelo Conoscenti, Martin Solly, Rachele Raus, Alessandra Molino for taking time to read previous versions of this book and providing helpful comments. Michelangelo was the first one who encouraged me to embark on this transdisciplinary journey and the one who endured supervising my work in the final stages. I would also like to thank Rachel Aldred, who kindly helped me to get access to the database from her ESRC project, and FIAB's Director Francesco Baroncini, who allowed me to attend an ECF Annual General Meeting. And my thanks go to Cathy Scott and Alice Green from Palgrave and to the anonymous reviewers for their important observations and suggestions.

I am also indebted to the many scholars who listened to my early thoughts on the subject of this book and provided suggestions, valuable opinions, different points of view and encouragement. Among them are Seraphim Alvanides, Esther Anaya, Johannes Angermuller, Osman Arrobbio, Jonathan Charteris-Black, Giulia Cortesi, Egidio Dansero, Francesco Gonella, Katja Leyendecker, Manfred Neun, Damien Ó Tuama, Dario Padovan, Paolo Ruffino, Andrea Scagni, Nadia Williams.

This book evolves from what I have learned as a cycling advocate and from the positivity and support I received from fellow activists. I thank all those who made our Bike Pride parades and our activities possible and, in particular, Elisa Gallo, Beppe Piras and Fabio Zanchetta, with whom I have often exchanged precious conversations.

Finally, I am grateful to my mum Jole for all her help and support and to Dario and Vanessa for making my everyday life so happy in so many ways. They are the main reason why I remain an incurable optimist.

Contents

List of Figures

List of Tables

CHAPTER 1

Introduction

Abstract This chapter introduces the framework of Ecolinguistics, which is employed in the book to bring together two fields that have rarely met before in academic literature: Cycling Mobility and Critical Discourse Studies (CDS). The ecosophy—or ecological philosophy—encompassing the analytical approach is outlined and explained, showing the centrality of life and the need to celebrate, respect and value all living beings. The chapter provides an overview of the sources of inspiration for the book and briefly explains the notions of Systems Thinking, Framing and Marketisation. The reasons for choosing cycling mobility as an object of investigation are introduced. The final part of this chapter describes the contents of the rest of the book and provides a brief summary of each chapter.

Keywords Ecosophy · Well-being · Systems thinking · Framing · Marketisation

The promotion of cycling as a solution to many of the problems affecting life in cities is increasingly present both as an academic object of study and in public debate and city planning. It is a topic that calls for an interdisciplinary approach, as the perspectives of Engineering and Architecture benefit from insights from Sociology, Geography, Behavioural Studies and Medicine—to name but a few. When surveying

M. C. Caimotto, *Discourses of Cycling, Road Users and Sustainability*, Postdisciplinary Studies in Discourse,
https://doi.org/10.1007/978-3-030-44026-8_1

the academic publications that investigate ways of increasing cycling and removing the existing barriers, it is possible to observe how discourse-related observations are frequent and prominent but, at the same time, how the field of Cycling Mobility is yet to become an object of study for Critical Discourse Studies (CDS). This book thus aims to bring together two different fields that have rarely met before in academic literature and to do so within the framework of Ecolinguistics.

The definition of Ecolinguistics we shall refer to is Stibbe's (2014, 217):

> questioning the stories that underpin our current unsustainable civilisation, exposing those stories that are clearly not working, that are leading to ecological destruction and social injustice, and finding new stories that work better in the conditions of the world that we face. These are not stories in the traditional sense of a narrative, however, but rather discourses, frames, metaphors and, in general, clusters of linguistic features that come together to covey particular worldviews.

Stibbe shows the connection linking Ecolinguistics and CDS, underlining how the discursive investigation of ecological matters expands what have traditionally been the objects of study of CDS, namely sexism, racism and other oppressive power relationships (see also Stibbe 2015).

Referring to these traditional objects of study, Halliday (2001) highlights that we have movements and commissions dealing with the role that language can play in reinforcing sexist and racist discrimination and attempting to change social reality by working on the language, but, on the contrary, we do not have anything similar regarding classism. Halliday explains:

> it is acceptable to show up sexism - as it is to show up racism - because to eliminate sexual and racial bias would pose no threat to the existing social order: capitalist society could thrive perfectly well without sex discrimination and without race discrimination. But it is not acceptable to show up classism, especially by objective linguistic analysis [...]; because capitalist society could not exist without discrimination between classes. Such work could, ultimately, threaten the existing order of society. (185)

Classism, as the analyses presented in this book will show, has a strong, deep, but not self-evident connection with mobility-related issues. This is revealed by Furness, who highlights that 'one of the cheapest forms of transportation on the planet is construed as elitist, whereas one of the most expensive and resource-intensive technologies is considered populist' (2010, 135). He observes the rhetorical device employed to construct cyclists as elitists, through the same sweeping critiques used against anti-car proponents and environmentalists. This, Furness argues, is a double-aimed conservative perspective employed to dismiss the questioning of the status quo and infer the existence of a silent majority who is against change. On top of this, it defines elitism as a disposition of moral or behavioural origin rather than something derived from privileges that are economic, structural legal or spatial.

Worldviews like the one exposed by Furness are observed in this book from an ecolinguistic perspective and their presence in the texts under scrutiny is exposed. Following Stibbe's (2014) proposal to strengthen the combination of CDS and Ecolinguistics, the investigations presented in this book judge worldviews on criteria that are derived from an ecosophy (or ecological philosophy) which brings together scientific evidence concerning the interaction of living beings, including humans, their physical environment and an ethical framework concerning why the survival and flourishing of ecosystems matters. This is in line with the Strong Programme in Discourse Studies outlined by Angermuller (2018, 6), which argues in favour of establishing the truth value of the claims under investigation by judging their normative quality (see Chapter 2). The worldviews investigated are judged from a scientific perspective and at the same time from an ethical perspective. More specifically this book aims to expose 'the ecologically destructive ways that everyday discourses construct notions of the "good life", providing tools to help resist those discourses, and searching for beneficial discourses which actively identify the "good life" with something other than consumerism' (Stibbe 2014, 125).

Stibbe (2015, 14–15) details the ecosophy used in his work and highlights the centrality of celebrating, respecting and valuing the life of all living beings. Living is more than 'being alive' and the author underlines the importance of well-being, explaining that it is a *sine qua non*, which has to be part of any measure to address ecological issues, given that any measure that harms human interests is not going to be adopted (14). The ability to live with high well-being should be granted to both present and future generations. The next central notion is that of empathy, which implies we need to be aware of the impacts of our choices, to minimise harm and feel a duty to give back to the system supporting us. This means, of course, that solutions to keep within environmental limits are essential, not exceeding in the consumption of resources or the creation of levels of waste that ecosystems cannot absorb. Social justice and resilience are the final points: as resources become exhausted and more ecological destruction is inevitable, solutions to redistribute resources among people and to find new forms of society are necessary in order to allow the continuation of high well-being.

The ecosophy on which this book is based on shares the above tenets and focuses on a specific aspect of life, everyday mobility, advocating the increase in the number of people cycling as an effective and significant solution that responds to all the necessities raised within the ecosophy. Increasing cycling numbers, especially in urban areas, means a significant reduction in vehicle-collision deaths and a reduction in sedentariness-induced deaths. It minimises the harm caused by the use of private cars in terms of fewer collisions, in terms of less consumption of resources—including the production of cars and car parts and, of course, the fossil energy required to run cars—and in terms of less production of waste, that is air pollution through exhaust fumes and waste from used car parts and cars.

Increasing cycling gives back to the system as people who cycle frequently report that cycling makes them appreciate more the surrounding environment and feel part of it (see Chapter 7), hence increasing cycle use is here assumed a way to start a positive chain-effect likely to generate higher levels of environmental awareness. As cycling is much more affordable than driving (not only in terms of economic costs, but also of the training and testing required), its increase also generates positive effects in terms of social justice. As for resilience, numerous studies show that the current levels of car use, projected in a medium-term future in which urban populations are forecast to grow steadily, would result in cities permanently locked in traffic congestions and increased cycling can ease this burden. It is important to point out that our focus on cycling is not meant to imply that no other solutions exist (e.g. walking, improving public transport, reducing the number of trips in the first place). Cycling was selected because it is the environmentally friendly mobility solution that appears most suited to replace private cars in urban environments, as it offers same or shorter journey times and same freedom from pre-planning (see Pucher and Buehler 2017), while generating health benefits thanks to the increase in physical activity. At the same time, cycling is the mobility choice that generates the highest levels of hostility from people who resist change.

Digging in the hegemonic underlying assumptions, that are found even in texts that aim to promote cycling, is a way to reveal the dominant destructive worldview within which all the above-mentioned destructive forces are at work. This is revealed by observing a topic, cycling-related discourse, which has been overlooked by CDS so far. The perspective from which this book was envisaged is also indebted to Meadows' book *Thinking in Systems* (2008). Leading author of 1972 *The Limits to Growth*, Meadows was a pioneering environmental scientist who dedicated her research to envisaging other ways to approach and tackle problems. With her background in chemistry and a PhD in biophysics, Meadows' work originates in disciplines that appear very distant from linguistics and discourse studies, but reveal deep and strong connections if we look closer. As Meadows (2008, 172) explains 'listen to any discussion, in your family or a committee meeting at work or among the pundits in the media, and watch people leap to solutions, usually solutions in "predict, control, or impose your will" mode, without having paid any attention to what the system is doing and why it's doing it'.

If we read her list of "places to intervene in a system" ranked in increasing order of effectiveness, the two top places are "Paradigms" and, even higher, "Transcending Paradigms" (2008, 162–165). Her definition of a society's paradigm is the 'deepest set of beliefs about how the world works. These beliefs are unstated because it is unnecessary to state them – everyone already knows them' (163). The examples of paradigm changes she mentions are Copernicus and Kepler showing that the earth is not the centre of the universe, Einstein's hypothesis that matter and energy are interchangeable and Adam Smith's postulation that 'the selfish actions of individual players in markets wonderfully accumulate to the common good', all situations in which the intervention hit a leverage point that totally transformed the system (163). These, of course, are all examples of changes in discourse, ideas that influence general knowledge so deeply that they generate new ways of thinking, new ways of talking and, consequently, new metaphors.

One of the most powerful messages emerging from Meadow's work is that while systems thinking is an approach that helps us look at a problem from a new and unexpected perspective, we must beware of the fact that systems thinking will not enable us to predict the future and that the implementation of the findings is much more problematic than retrieving the findings in the first place. She thus recommends to 'dance with the system', getting its beat, listening to its wisdom, celebrating complexity (see Meadows 2019 for the full list of places to intervene). Meadows (2008, 175–177) underlines the importance of quality over quantity, highlighting how many of our attempts to solve problems by changing something in the systems are based on quantitative information, for the simple fact that quantities are easier to measure and to employ as proof. This strategy eventually will inevitably create systems that attribute more prominence to quantity rather than quality (the notion of measuring the performances of nations through the GDP being the most blatant example of this mistake). Meadows underlines the importance of focusing on values rather than numbers, a piece of advice that corresponds to that provided by Lakoff, who recommends to 'talk at the level of values and frame issues in terms of moral values' (2010, 79).

Lakoff (2010) and Mautner (2010) are at the centre of Chapter 2 as their works were the first sources of inspiration for this research. Chapters 2 and 3 provide the background for the analyses presented in the book. The former outlines the key concepts and briefly explains the theoretical approach. It introduces CDS and describes the publications that have inspired this investigation, providing an overview of how CDS can be employed to investigate the discourse of everyday cycling and why it is a worth undertaking. The notion of hierarchy between different systems, which is fundamental in systems thinking (see also Sterman 2012; Gonella et al. 2019), is present in the work of Mautner, who explains how, starting from the neoliberal era of 1980s, the economy has been envisaged as the top system in the hierarchy, embedding society rather than the opposite (2010, 16). In her work, Meadows (2008, 178) underlines that a common mistake is to misunderstand and misplace the hierarchy governing the various systems. The ongoing climate crisis is a consequence of the dominant trend that sees the market and its logic as the overarching system that encapsulates everything else. But the core problem of the climate crisis is that the market requires resources that are finite, and thus it is the planet from which such resources are extracted that consists in the overarching system within which the others are embedded. Chapter 2 surveys other examples of metaphorical thinking and their effects and exposes the rationale that explains the choice of cycling as the object of investigation.

Chapter 3 aims to fill the gap for readers who have a linguistic background and may not be familiar with the history, challenges and goals of promoting the growth of everyday cycling. Studies from other disciplines are reported, bringing together what needs to be known in order to understand the issues at stake. The history of the bicycle is briefly sketched, pointing out how it has subsequently been envisaged as a pseudo-Darwinist evolution (Cox and van De Walle 2007). The growth of the automobile system, the introduction of the idea of 'jaywalking' (Norton 2007) together with the impact of the work of Colin Buchanan (1963–2015) show the changes in mentality that contributed to the creation of automobility as we know it. This is observed against the wider perspective of the notion of sustainability, and how it has become intertwined with that of economic development (Swyngedouw 2007). Wild et al.'s (2018) work on 'bikelash' is reported, as it helps understand the various sources of opposition any urban project is likely to face when introducing cycling-friendly infrastructure. Urry's (2004) notion of car system and his investigation into the implications of automobility contribute to the understanding of the wider picture. The chapter ends with a review of the controversial arguments that are typically raised when discussing cycling in urban environments, that is, the perception of ownership of the urban space, mandatory helmet laws, the Safety in Numbers effect and the exposure to particulate for people who cycle.

Chapters 4–7 present linguistic investigations of a variety of case studies, aiming to provide an overview of some approaches that ecolinguists can follow when investigating discourses of mobility and sustainability. The four case studies open with the investigation of existing narratives of conflict and move to the observation of the presence of hegemonic marketised discourse. The texts under scrutiny are analysed in order to retrieve 'us vs. them' narratives together with the discourse strategies deployed to avoid reification and conflict. The institutional texts are also observed to detect the presence of hegemonic marketised discourse, reflecting on the consequence of its presence. The interview transcripts are then employed to identify other narratives that are less frequently employed in institutional texts promoting more cycling.

The first case study, presented in Chapter 4, consists in a fine-grained qualitative analysis of a selection of articles published in British national newspapers concerning the trial ensued from a collision between a cyclist, Charlie Alliston, and a pedestrian, Kim Briggs, which happened in London in February 2016, causing the subsequent death of Kim Briggs. The analysis is carried out by applying the discursive moves identified by van Dijk (1992) to observe the denial of racism in the press. This case study was selected because the tragic outcome of the collision-lowered filters and the texts analysed to show how cyclists, and even more cycling advocates, are constructed as the enemy and are conceived as one group which—as Furness (2010, 131) argues—is held collectively responsible for the misdemeanour of one person. These observations are taken into account from a wider perspective, showing the links between racism and discrimination against people who cycle, envisaging them as two irrational and exaggerate reactions stemming from fears of change.

In Chapter 5 the focus is on institutional discourse and, while Chapter 4 investigates mainly the 'us vs. them' perspective, the analysis of documents published by the Mayor of London and Transport for London in 2018 shows the coexistence of both threads, highlighting that on the one hand the authors took great care to avoid conflictual language and focused on the notion of citizens rather than dividing them into road user categories but, on the other hand, they subscribed to the dominant ideology of 'growthism' (Halliday 2001). In this chapter, some basic notions of discourse analysis and corpus-aided methods are explained to help readers who do not have a background in linguistics. These are then applied to the analysis of the texts in order to reveal underlying discourses, identifying which ones contribute to a change in cycling-related discourse and which ones perpetuate the same conceptual mistakes that have led humanity to the current climate crisis.

With the investigations presented in Chapter 6, the focus moves beyond Great Britain to analyse texts and tools envisaged mainly for a European readership. These are the European Cyclists' Federation's 'EU Cycling Strategy' (ECF 2017); the 'Declaration on Cycling as a climate friendly Transport Mode' (EU 2015); the 'Graz Declaration' (EU 2018) signed by European Environment and Transport Ministers; the white paper 'Sustainable Urban Transportation, Creating Green Liveable Cities' (SOG 2016) published by the Danish public–private partnership State of Green and the HEAT tool developed by the World Health Organization (WHO) (Kahlmeier et al. 2010). The chapter opens with the history of the notion of 'sustainable development' and describes its paradoxes, bearing in mind the widespread mistakes concerning the system hierarchy. The investigations, through a corpus-aided approach, reveal the presence of marketisation in texts aiming to promote an increase in cycling levels and offers some reflections on the dangers that an economy-focused discursive strategy can entail.

Discourses of cycling and road users are observed from a different viewpoint in Chapter 7. This time, the texts under scrutiny are transcriptions of spoken discourse, retrieved from the interviews which

Aldred and her team carried out as part of the *Cycling Cultures in a Mass Motorised Society* ESRC project (see Aldred 2012a, b, 2015). The investigations, once again through a corpus-aided qualitative approach, reveal strategies of identity construction and framing in the words of people who cycle regularly but, in most cases, are not 'hardcore' cyclists. The aim of this investigation is double, as it reveals the existing frames that hinder increases in cycling numbers while it also retrieves frames and viewpoints of people who cycle that can be employed to construct new effective ways of promoting cycling.

These new ways are outlined and described in Chapter 8. Moving beyond the anti-neoliberalist perspective against which Purcell (2016) warns us, this chapter focuses on positive discourse strategies. Suggestions that can be employed as starting points to promote cycling are presented and explained, drawing on the positive examples observed in the case studies investigated in the book. In line with Stibbe's proposal exposed above, the concluding chapter highlights the importance of celebrating, respecting and valuing the life of all living beings; it also presents the starting points from which new worldviews and beneficial discourses can be framed and promoted, talking on the level of values and changing the paradigms of the system.

References

Aldred R. (2012a). *Cycling Cultures: Summary of Key Findings and Recommendations.* https://westminsterresearch.westminster.ac.uk/item/8z5y3/cycling-cultures-summary-of-key-findings-and-recommendations. Accessed 2 June 2019.

Aldred, R. (2012b). *Cycling Cultures in a Mass Motorised Society* [computer file]. Colchester, Essex: UK Data Archive [distributor], September 2014. SN: 7545. http://dx.doi.org/10.5255/UKDA-SN-7545-1.

Aldred, R. (2015). A Matter of Utility? Rationalising Cycling, Cycling Rationalities. *Mobilities, 10*(5), 686–705.

Angermuller, J. (2018). Truth After Post-truth: For a Strong Programme in Discourse Studies. *Palgrave Communications, 4*(30), 1–8.

Buchanan C. (1963–2015) *Traffic in Towns—A Study of the Long Term Problem of Traffic in Urban Areas. Introduction by Simon Gunn.* London and New York: Routledge.

Cox, P., & Van De Walle, F. (2007). Bicycles Don't Evolve: Velomobiles and the Modelling of Transport Technologies. In D. Horton, P. Rosen, & P. Cox (Eds.), *Cycling and Society* (pp. 113–132). London: Routledge.

ECF. (2017). *EU Cycling Strategy: Recommendations for Delivering Green Growth*. https://ecf.com/eu_cycling_strategy. Accessed 10 June 2019.

EU. (2015). *Declaration on Cycling as a Climate Friendly Transport Mode*. http://www.eu2015lu.eu/en/actualites/communiques/2015/10/07-info-transports-declaration-velo/07-Info-Transport-Declaration-of-Luxembourg-on-Cycling-as-a-climate-friendly-Transport-Mode—2015-10-06.pdf. Accessed 10 June 2019.

EU. (2018). *Graz Declaration*. https://www.eu2018.at/latest-news/news/10-30-Graz-Declaration.html. Accessed 10 June 2019.

Furness, Z. (2010). *One Less Car: Bicycling and the Politics of Automobility*. Philadelphia: Temple University Press.

Gonella, F., Almeida, C. M. V. B., Fiorentino, G., Handayani, K., Spano, F., Testoni, R., & Zucaro, A. (2019). Is Technology Optimism Justified? A Discussion Towards a Comprehensive Narrative. *Journal of Cleaner Production, 223*, 456–465.

Halliday, M. A. K. (2001). New Ways of Meaning: The Challenge to Applied Linguistics. In A. Fill & P. Mühlhäusler (Eds.), *The Ecolinguistics Reader: Language, Ecology and Environment* (pp. 175–202). London and New York: Continuum.

Kahlmeier, S., Racioppi, F., Cavill, N., Rutter, H., & Oja, P. (2010). "Health in All Policies" in Practice: Guidance and Tools to Quantifying the Health Effects of Cycling and Walking. *Journal of Physical Activity and Health, 7*(1), 120–125.

Lakoff, G. (2010). Why It Matters How We Frame the Environment. *Environmental Communication, 4*(1), 70–81.

Mautner, G. (2010). *Language and the Market Society: Critical Reflections on Discourse and Dominance*. London: Routledge.

Meadows, D. H. (2008). *Thinking in Systems: A Primer*. Diana Wright (Ed.). White River Junction, VT: Chelsea Green Publishing.

Meadows, D. H. (2019). Dancing with Systems. *The Donella Meadows Project: Academy for Systems Change*. http://donellameadows.org/archives/dancing-with-systems/. Accessed 8 November 2019.

Meadows, D. H., Meadows, D. L., Randers, J., & Behrens, W. W., III. (1972). *The Limits to Growth: A Report for the Club of Rome's Project on the Predicament of Mankind*. New York: Universe Books.

Norton, P. D. (2007). Street Rivals: Jaywalking and the Invention of the Motor Age Street. *Technology and Culture, 48*(2), 331–359.

Pucher, J., & Buehler, R. (2017). Cycling Towards a More Sustainable Transport Future. *Transport Reviews, 37*(6), 689–694.

Purcell, M. (2016). Our New Arms. In S. Springer, K. Birch, & J. MacLeavy (Eds.), *The Handbook of Neoliberalism* (pp. 613–620). London and New York: Routledge.

SOG. (2016). *Sustainable Urban Transportation: Creating Green Liveable Cities*. https://stateofgreen.com/en/uploads/2016/06/Sustainable-Urban-Transportation.pdf?time=1565032648. Accessed 10 June 2019.

Sterman, J. D. (2012). Sustaining Sustainability: Creating a Systems Science in a Fragmented Academy and Polarized World. In M. P. Weinstein & R. E. Turner (Eds.), *Sustainability Science: The Emerging Paradigm and the Urban Environment*. New York: Springer.

Stibbe, A. (2014). An Ecolinguistic Approach to Critical Discourse Studies. *Critical Discourse Studies, 11*(1), 117–128.

Stibbe, A. (2015). *Ecolinguistics: Language, Ecology and Stories We Live by*. New York and London: Routledge.

Swyngedouw, E. (2007). Impossible "Sustainability" and the Postpolitical Condition. In R. Krueger & D. Gibbs (Eds.), *The Sustainable Development Paradox: Urban Political Economy in the United States and Europe*. New York and London: Guilford Press.

Urry, J. (2004). The 'System' of Automobility. *Theory, Culture & Society, 21*(4–5), 25–39.

van Dijk, T. A. (1992). Discourse and the Denial of Racism. *Discourse and Society, 3*, 87–118.

Wild, K., Woodward, A., Field, A., & Macmillan, A. (2018). Beyond 'Bikelash': Engaging with Community Opposition to Cycle Lanes. *Mobilities, 13*(4), 505–519.

CHAPTER 2

Discourse Practices and Power

Abstract This chapter outlines the main Linguistics publications that have inspired, supported and driven the perspectives from which the case studies analysed in the book are investigated. It also draws from studies belonging to other disciplines in order to explain why an investigation into the discourse of the promotion of everyday cycling from a linguistic perspective can provide fruitful insights. The notion of discourse and a brief history of the school of Critical Discourse Studies (CDS) is explained. The chapter focuses on what connects (re)framing the environment and the hegemonic tendency towards marketisation in language. The centrality of metaphors is highlighted, showing that we *think* through metaphors by referring to studies from other disciplines and in particular to the very first economic theories. The chapter also introduces the notion of dehumanisation, reporting a few studies that have applied it to people who cycle.

Keywords Metaphorical thinking · Dehumanisation · Critical Discourse Studies · Semiotics · Economic theories

This chapter outlines the main Linguistics publications that have inspired, supported and driven the perspectives from which the case studies analysed are investigated. It also draws from studies belonging to other disciplines in order to explain why an investigation into the discourse of the

M. C. Caimotto, *Discourses of Cycling, Road Users and Sustainability*, Postdisciplinary Studies in Discourse,
https://doi.org/10.1007/978-3-030-44026-8_2

promotion of everyday[1] cycling from a linguistic perspective can provide fruitful insights. As already mentioned, this work follows a critical tradition and aims to reveal the discourses lying under the surface of texts dealing with everyday cycling, whether in order to promote it or to hinder it. Before proceeding, a definition of discourse is likely to prove useful. 'Discourse' can have many definitions—as these can vary across the various disciplines that use the term—and its meaning has been stretched from a genre to a register or style, causing much confusion (Wodak and Meyer 2016, 4). Angermuller (2015) provides an overview of the various discourse theories and approaches that have informed the cross-disciplinary perspective that exists today and explains that 'what makes discourse such a methodological challenge is that it is not a pure object. [...] Therefore, the heterogeneous constitution of a discourse needs to be emphasized, which consists of different but interwoven "components": "language", "practice" and "context"' (2015, 512).

In Critical Discourse Studies (CDS) discourse consists in a particular representation of the world. According to a definition provided by Fairclough and Wodak (1997, 258), scholars who identify with CDS see discourse as a form of social practice. This implies that a particular discursive event is shaped by the situation(s), institution(s) and social structures that frame it but at the same time the discursive event also shapes them, in a dialectical relationship. Discourse can help to sustain and reproduce the situations, objects of knowledge, social identities and relationships between people or groups of people, but it can also help their transformation. Discursive practices thus have important consequences for society at large and give rise to important power-related issues, as they can help produce and reproduce unequal power relations through the ways in which they position people and represent things (Fairclough and Wodak 1997, 258).

This notion of discourse as linked to power has its origins in the work of Foucault (1971): the important aspect he introduced in his theory is that the ways in which we use language play an active role in the construction of reality as discourses shape and enable (social) reality (see also Angermuller et al. 2014, 98–110). Discourse is a material reality of its own that cannot be reduced to language merely referring to objects that

[1] I use the phrase 'everyday cycling' as an alternative to the labels 'utilitarian cycling' or 'utility cycling', as the problematic aspects of the label 'utilitarian/utility' will be discussed further on in the book.

exist in reality or some form of 'false consciousness' or 'distorted view of reality'. It is through discourses that ways of talking, thinking and acting are institutionalised and regulated, and it is in this sense that discourses exercise power (see Fairclough 1992; Jäger and Maier 2016). CDS focus on what Fairclough (2016, 91) calls 'social wrongs', that is forms or orders which are detrimental to human well-being, situations in which unequal power relations are favoured and sustained by discursive practices. The main social issues investigated by CDS scholars concern racism, sexism and other forms of discrimination based on religion or sexual orientations, but more in general CDS is interested in simplistic ways of representing the world that create an 'us' vs. 'them' narrative.

The fact that discourses exercise power over people could be interpreted as a negation of the individual, but it is important to stress that this way of conceiving discourse should not be interpreted as a denial of agency, the subjects involved in the production and reproduction of discourses are not passive means through which discourses exert power. On the contrary, individual subjects collectively play a role in conserving or changing the dominant relationships of power at work in a specific place at a specific time. In fact, Fairclough specifies that 'discourse is a mode of action, one form in which people may act upon the world and especially upon each other' (1992, 63). Individuals who have privileged access to education, knowledge and wealth have more power to exert their influence over the dominant discourses at a given time and place, but no individual or group can defy hegemonic discourse or have full control over discourse.

Influence—whether aiming to reproduce or contest the dominant discourse(s)—will not be exerted through a single text but rather through constant repetition. Changes are thus generated either slowly, through a gradual shift in the ways reality is discussed and perceived, or more swiftly in the wake of some major historical event able to shake people's deepest convictions. The prevailing social circumstances—for example the power relationships at work in a specific historical moment—will determine whether the language will reproduce the existing power structures, manufacturing consent or rather challenge them and sustain resistance. Fairclough (1992, 65) points out that it is important to avoid the pitfalls of overemphasising on the one hand the idea that discourse is a mere reflection of a deeper social reality and on the other hand the construction of the social in discourse: consequently the relationship between discourse and social structure should be seen dialectically.

In fact, Angermuller (2018, 2) explains, discourse theorists have been criticised for having excessively questioned reality vis à vis the mass media society and have been blamed for discrediting the idea of scientific truth. In order to overcome the risk of discourse research becoming relativist rather than constructivist (i.e. asking how ideas are constructed practically), Angermuller outlines a Strong Programme in Discourse Studies, explaining that it is possible to acknowledge that some truth claims have more truth value than others by judging their normative quality, without betraying the constructivist approach: 'rather than deciding on what is the one true perspective on the social, [the Strong Programme] takes stock of the various, competing truths and realities and ask how they are constructed' (Angermuller 2018, 6).

The school of CDS originated in the early 1990s and has always avoided the creation of a 'closed' theory or a specific method (see van Dijk 2013; Angermuller et al. 2014). It rather sees itself as a heterogeneous school bringing together various and varied methodological and theoretical approaches, while scholars who identify with it share the perspective from which they approach the analysis of discourse. As Bloor and Bloor (2007, 12) aptly describe with a metaphor, the aim of CDS researchers is 'to investigate how ideologies can become frozen in language and find ways to break the ice'. 'Ideology' is another concept that defies being captured easily, as most discourse-oriented studies understand ideology as *a coherent and relatively stable set of beliefs or values* (Wodak and Meyer 2016, 8) but it is difficult to separate this way of viewing it from the negative connotations attached to it in everyday language and in the work of some scholars.

The perspective from which this investigation into the discourse of cycling is carried out was inspired mainly by two works: Mautner's book *Language and the Market Society* (2010) and Lakoff's paper *Why It Matters How We Frame the Environment* (2010). Both published in the wake of the 2008 recession, these works reveal the discursive presence of the all-pervasive paradigm of the market society. They explain why, in different contexts, the marketisation of discourse is toxic and prevents people from achieving the goals they are pursuing because such goals are envisaged, discussed and understood through a prism—the market—which is unrelated to them and hinders the real and full understanding of the issues at stake. As a response to their findings, following Stibbe's precious observation, this work will attempt to 'search for new discourses to base society on; for example discourses that promote *being more* rather than *having more*, well-being rather than growth and respecting rather than conquering nature' (2018, 165). The four case studies are analysed applying both negative and positive discourse analysis (Martin 2004) in order to draft new ways of narrating, framing and envisaging the discourse of cycling.

Lakoff's analysis highlights the difficulties that environmentalism faces as a consequence of a series of tenets that, he explains, do not hold. Firstly, there is what Lakoff calls 'the Enlightenment Reason', the idea that 'if you just tell people the facts, they will reason to the right conclusion' while, what actually happens is that the facts must make sense in the recipients' systems of frames, otherwise they will be rejected and ignored (2010, 72–73). People's systems of frames consist in unconscious structures that allow us to think, they include 'semantic roles, relations between roles and relations to other frames' (ibid., 71). As these structures physically take the form of neural circuits in our brains, every word we use evokes a frame and, as a consequence, every situation in which we think and talk involves 'framing'. Frames include everything a person knows about something and what they associate to that, every word included in the frame will evoke the frame. If the activity of a surgeon is being described, for example, 'operating theatre, scalpel, nurse, patient' are evoked and the sentence 'the patient cut the surgeon's staples' does not fit the frame. Many of these frame-circuits are directly connected to the emotional creation taking place in our brains, which means that thinking involves emotions—a view that is obviously very distant from the idea that by thinking rationally and focusing on facts we can take rational decisions, and avoid being influenced by emotions.

This way of conceiving how we think challenges the dominant ways in which knowledge and understanding are conceived and understood and has only been developed in the last 25 years, this explains why we still fall into 'the trap of the Enlightenment Reason'. Lakoff (2010, 77) goes on to explain that when talking about the environment we suffer from hypocognition, as we lack the frames we need to understand the very messages we are discussing. The first and most evident problem is that the environment is conceived as external from ourselves, we talk about the environment as something from which we are separate, while in fact as human beings we cannot exist without the nature of which we are an intrinsic part. This division is so deeply ingrained in our brains that it is particularly difficult to remove it from our way of thinking.

While we consider ourselves detached from the environment, when it comes to our ways of conceiving humanity and the social relationships we are engaged in, we increasingly accept to view any aspect of our life through the filter of monetarised exchange. As Mautner explains, the market and the need to increase 'the bottom line'—a company's net income—are increasingly becoming the social principle driving any human relationship. What is being criticised, and presented as problematic, is not the presence of a market per se or the fact that market exchange is driven by a marketing logic, but that—starting from the late-modern, neoliberal climate of the 1980s—the society has been embedded within the economy rather than the opposite (Mautner 2010, 16). This 'marketisation' consists in the transfer of the laws of the marketplace to worlds that were not originally organised following a market logic, and its effects are visible both in social practice and in discourse, giving rise to *marketised discourse*.

Mautner underlines the difference between *marketised discourse* and *market discourse*, pointing out that talking about sales in a shop consists in *market discourse*, while a sermon made to sound like sales talk is *marketised discourse*, the difference ultimately lying in the respect for boundaries, as marketised discourse takes the market discourse beyond what should be its limits (2010, 17). The way in which marketised discourse manifests itself in the language we speak is through a constant use of metaphors drawn from the realm of the market economy. We are so used to their presence that we do not really notice them anymore, as on the contrary, we would probably do if religious or military metaphors were to dominate our communication. Of course, Mautner argues (30–31), borrowings from other domains such as the military occur in our language, but they are marked, they catch our attention. They may be used with support, or ironically or disapprovingly, while marketised discourse goes unnoticed and is broadly perceived as natural.

Marketisation is in fact a powerful conceptual metaphor. A conceptual metaphor is a metaphor through which we understand the world around us and with which we *think* about aspects of the world. According to Lakoff and Johnson (2003, 156):

> Metaphors have entailments through which they highlight and make coherent certain aspects of our experience.
>
> A given metaphor may be the only way to highlight and coherently organize exactly those aspects of our experience.

> Metaphors may create realities for us, especially social realities. A metaphor may thus be a guide for future action. Such actions will, of course, fit the metaphor. This will, in turn, reinforce the power of the metaphor to make experience coherent. In this sense metaphors can be self-fulfilling prophecies.

We can see here another connection between the work of Mautner (2010) and that of Lakoff (2010), the fact that the more we use marketisation as a way of interpreting how we conduct our lives and how we interact with the environment of which we are part, the more marketisation will become powerful. This process can lead to a domination of the metaphors of marketisation. Metaphors are the main discursive strategy through which framing is enacted, as the recipients interpret the link being created between the source domain (the concept we draw upon in order to create the metaphor) and the target domain (the topic or concept that we want to describe through the metaphor) (Lakoff and Johnson 2003, 253). Recipients thus think of the target domain in terms of the source domain and this affects their understanding of the target domain itself. While a metaphor can be fruitfully employed to simplify an abstract and complex notion, if the source domain selected is not apt to describe the target domain, then the target domain will be misunderstood and the outcome will be hypocognition, that is the inability to understand the issues at stake because of a lack of adequate frames through which they can be interpreted.

Ironically, according to some prominent economists (Stiglitz 2004; Rifkin 2011) the dominant strands of thinking in economics—theorised by Adam Smith, Jean Baptiste Say and others—evolved from the use of metaphors drawing upon Newtonian physics and Darwin's evolutionary biology, because these were the dominant scientific discoveries when the first economic theories were introduced and economists referred to physics and biology to give credibility to their findings. But, Rifkin explains, the use of these metaphors actually gives us a false sense of how economic activity unfolds. This metaphorical way of thinking, together with the teleological interpretation of evolution—which was not present in Darwin's own work (Stiglitz 2004, 23–24)—have often been employed to justify market liberalism: if only the fittest survive, then society will become increasingly fit if the invisible hand of the market is let free to act.

In fact, Stiglitz shows how the transformations brought by globalisation and the New Economy do not leave everyone better off and he convincingly argues that we should rather 'structure economic and political processes in ways that make it more likely that the changes we face will in fact constitute meaningful progress' (2004, 25). According to Rifkin, only in the second half of the nineteenth century, with the theorisation of the first and second laws of thermodynamics, economists had a scientific basis they could employ to describe how economic activity works, but the metaphors based on Newton's work were so dominant in their theoretical thinking by then, that they refrained from abandoning them even if they were largely inapplicable to the economic sector (2011, 195).

All this is a confirmation of two points made by Lakoff (2010): metaphorical thinking is so dominant and powerful that it can actually lead to misinterpreting the reality around us because we are only able to understand it if the interpretation makes sense with our system of frames and changing the system of frames requires a long time. The fact that the laws of thermodynamics, the laws of energy that shape every single aspect of life as we know it, are better suited to be the scientific basis through which we could understand how the economy works underlines once more how we are much more deeply embedded in the natural environment than we think, as we prefer to conceive the environment as something that surrounds us but from which we are distinct and separate. The effects of climate change are a constant reminder of how we are part of the environment: climate change is a phenomenon which anthropogenic origins are now widely recognised and linked to the process of industrialisation, its existence affects the well-being of humans through many macro and micro effects and also menaces the survival of humanity.

The investigation presented in this book focuses on a single specific aspect of social change which could contribute to the improvement of people's well-being: the increase of everyday cycling, which is the increase of the number of people who use a bicycle as their main means of transport for their everyday mobility needs. A numerically relevant shift in mobility choice from private cars to bicycles could result in a significant reduction in air pollution, together with a number of positive effects on people's health. Of course, also a shift from private cars to walking would have similar effects, and, less significantly, a shift from public transport to active mobility. Everyday cycling is an example of how the promotion of specific policies to generate a change in social habits concerning mobility choices would generate positive effects on people choosing to cycle (see Raser et al. 2018) and—at the same time—to the environment and to any other person dwelling in the same environment, thanks to the improved air quality, and many other positive effects that will be discussed more in detail in the next chapter. While a growing number of sociologists, geographers, psychologists and political scientists are adding their contribution to the scholarly research aiming to reveal the existing barriers that prevent this desirable increase from taking place, everyday cycling has not yet been investigated from a discursive and linguistic perspective—the only partial exception being Kiernan's (2018) work, in which Systemic Functional Linguistics and Multimodal Analysis are employed to examine how identities associated with cycling are narrated in digital environments. There are several reasons why I believe that investigating the discourse of cycling from linguistic-discursive perspective is a worth undertaking.

Discourse concerning cycling involves power struggles in a realm that CDS have not yet taken into account, at least from a linguistic perspective. If we think about the most traditional realms investigated by CDS—notably sexist, racist and religious discrimination—the people who usually occupy a prominent role in our societies are white, middle or upper-class men. Men belonging to this category constitute the main percentage of the people identified and reified as 'cyclists' and represent the weak part in the relationship with other road users and, in particular, drivers. Of course, a cyclist's identity—being simply based on the means of transport the person is riding—is more volatile and transient than identity tied to gender, nationality or religion. The role played by discourse in construing the reality of cyclists as a community is thus particularly relevant and fascinating. It is important to underline that, as is the case with any other form of discrimination, the imagined community of 'cyclists' is distant from the actual reality of people who cycle, which is why in this book the two phrases are employed to distinguish the stereotyped community (cyclists) from the actual people who use a bicycle (people who cycle) as one of their means of transport. Andrews et al. (2018) observe how the underdevelopment of narratives around disabled cyclists contribute to restricted perception of which people are expected to cycle and argue in favour of measures to improve their visibility through imagery and language.

The relevance of a linguistics perspective is also tied to the fact that the domination of one means of transport—the private car—over the others is constructed discursively both through language—in the media, in advertising, in political discourse, in everyday conversations—and through the semiotics of road signs and the organisation of urban spaces. And it is also discursively that the hostility towards cyclists or towards the creation of cycle-friendly infrastructures are enacted and reinforced. On the other hand, the promotion of cycling is enacted through persuasive discursive strategies, both in the form of advocacy discourse to convince administrators at various levels to implement solutions that will increase cycling and in the promotion of cycling by institutions, or by grassroots associations and individuals to convince other people to modify their mobility choices and include cycling, or increase their bicycle use. When administrators decide to introduce or improve the infrastructures to make cycling safer and more attractive, they do so by changing the semiotics of urban space and, sometimes, by using persuasive strategies, such as advertising billboards, to convince more people to cycle.

In order to understand the actual results of policies promoting cycling, Oosterhuis (2019) carried out an extensive meta-analysis on more than two hundred published and unpublished research papers and several policy documents produced in the last two decades, concerning towns and cities in Europe, North America and Australia. His conclusions, if probably somewhat depressing for activists and pro-cycling policymakers, are relevant to underline the need for a new perspective and new approaches. He states that 'policies during the last two decades have largely failed to generate significant increases in everyday cycling in countries with low to average pedalling levels, whereas in countries with relatively high cycling volumes, such policies have contributed to a consolidation of its existing, relatively high level, rather than to further growth' and points out that 'it may be wise to shift the focus in bicycle policies from rational planning to nudging[2] strategies in order to influence through more subtle, socio-psychological and cultural means the engrained habits and attitudes that play such a crucial, but not always clearly visible motivational role in traffic behaviour and mobility patterns' (2019, 95–97).

We thus find the same problem revealed on a large scale by Lakoff, when he explains that people will not act to tackle climate change and will not change their habits on a rational basis simply because they are exposed to scientific facts (2010, 72), and on a smaller scale by

[2]In Chapter 5 we shall discuss why the notion of "nudging" is controversial (see Mulderrig 2018).

Oosterhuis, who explains that *habitus*—'ingrained and taken for granted collective habits'—is much more influent when it comes to changing one's habits than individual and rational-instrumental consideration (2019, 53). As both authors underline, changing the frames, the ways in which people conceive the issues at stake, requires a long time and is not simple nor straightforward. But this does not diminish the importance of undertaking the challenge, as creating effective communication that does reach people's attention and can affect their choices is essential in order to obtain significant results within a realm that is becoming increasingly urgent and can decide the survival or the extinction of human life as we know it. The investigation into the discourse of cycling presented in this book aims to bring to the surface underlying currents, observe what can hinder and what can actually help to promote a change in people's cognitive approach to issues related to the environment and people's well-being.

Of course, changing the mobility habits of a significant number of individuals will not solve the issues of climate change by itself and it is not the panacea for all our environmental problems. The reason why I believe the discourse of cycling is worth investigating is that it embeds issues of social power and conflict that could reveal patterns and frames which are at work in other realms and are hindering the ways in which we perceive and respond to the challenges posed by climate change. As both Sheller (2018) and Halliday (2001) point out, environment-related problems are deeply linked to problems of social justice. Given that the activity of cycling (even if not necessarily everyday cycling) is shared and appreciated by people who belong to any social class, ethnic group, gender, age, religion and who have political views that cover the whole possible spectrum, perhaps the promotion of everyday cycling, if well framed, could prove a powerful tool to persuade people to care more about environmental issues and to deconstruct, or at least weaken, the division ingrained in our brains that we are detached from the environment (Lakoff 2010, 77), highlighting at the same time that respecting the environment and respecting fellow humans are mutually dependent choices.

One of the aims of the analysis presented in this book is to investigate the presence of marketisation in the discourses about cycling, observing when the promotion of everyday cycling reveals traces of marketisation and reflecting on the potential effects of this contamination of discourses. At the same time, cycling and, in particular, the bicycle as an object, is itself often employed as a metaphor, especially in advertising. These images of cycling and bicycles are often employed to entail positive associations (e.g. love, freedom, happiness) that advertisers employ to promote all kinds of products and services that are totally unrelated to cycling. While these images can reinforce positive association to cycling in our brains, they also reinforce the idea that our happiness and freedom depend on the products and services we decide to buy, which is of course one of the effects of marketisation.

As with any other form of discourse, language shapes and is shaped by society, hence the more marketised discourse spreads, the stronger it gets. And in a spiralling effect, organisations of all kinds will be increasingly attracted towards it, they will embed marketised discourse in their communication and will thus reinforce its appeal. In her work, Mautner (2010) shows the influence of marketised discourse on spheres that have little to do with market forces: the public-sector administration, higher education, religion and the personal sphere. She does not investigate environmental discourse but her analysis connects well with Lakoff's observations: he convincingly argues that it is difficult to discuss environmental issues effectively and to find solutions because there is no established frame in public discourse for the deep truth that '*the economic and ecological meltdowns have the same cause*, namely, the unregulated free market with the idea that greed is good and that the natural world is a resource for short-term private enrichment' (2010, 77 emphasis in the original).

While it is not yet part of public discourse, there is a new economic approach that focuses on well-being rather than growth and aims to find new ways of measuring progress in order to abandon the idea that GDP (Gross Domestic Product) is the ultimate way of establishing whether a country is improving or not (see Kubiszewski et al. 2013; Stiglitz et al. 2010). These new indexes (e.g. GPI Genuine Progress Indicator) are not limited to the observation of the economic aspects but take into account also environmental and social elements. The mistake of using GDP as a broader measure of welfare is exemplified by the fact that an oil spill generates an increase in GDP (as a consequence of the costs of clean up) but spilling oil is certainly not a way to increase people's well-being or environmental quality. Thus, from various perspectives, the domination of marketised discourse over all aspects of human life, and well beyond the boundaries of issues that belong to the realm of the market, is one of the main forces that hinder the possibility of conceiving new ways of organising our lives and enact the necessary changes that will allow the survival of the ecosystems on which also human life depends.

Two main threads are employed in this book to investigate cycling-related discourse, one is that of marketisation, which is demonstrated by many scholars across various disciplines to be the dominant paradigm through which our societies are organised. The other is the presence of as 'us vs. them' narrative, which is also a widespread and common phenomenon at work in a vast number of social issues, but one that has not yet been observed through a linguistic lens in cycling-related discourse. A number of studies, mainly from a psychological perspective, have investigated the attitudes of drivers towards cyclists and the overall findings show that the associations that drivers make are predominantly negative rather than positive (Basford et al. 2002). But while other forms of discrimination, if stated publicly and openly, will be received with outrage, the widespread perception and reaction to discriminatory statements against cyclists is less negative. As Johnson points out, examples of public reference to violence against cyclists—such as aggressive statements in the style of 'all cyclists should be killed' made by public figures—are not uncommon but they are rarely condemned as would be, for example, violence towards women or bullying. In fact, she underlines, it is hard to think of another group engaged in a legal activity against whom such levels of threat with violence and hostility would be considered acceptable (Johnson 2014).

A recent pilot study (Delbosc et al. 2019) has investigated attitudes towards people who cycle by applying 'dehumanisation measures' within the transport sphere for the first time. Dehumanisation is usually understood as a phenomenon that takes place in extreme situations such as genocide or war (Lang 2017), but it simply refers to situations in which a human being is considered and treated as 'less than fully human' (Haslam and Stratemeyer 2016, 25). Delbosc et al. found that 'a significant proportion of the survey sample dehumanized cyclists'. They warn that their study is only a pilot and their sample over-represents young, high-income males. Still, they explain that their findings open important new patterns of research, as it shows that dehumanisation measures can be fruitfully employed to investigate the nature of aggression towards cyclists. Drawing upon previous research that focused on different examples of dehumanisation (immigration, for example), they explain that dehumanisation attitudes can predict later aggressive behaviour and hypothesise that 'dehumanizing attitudes toward cyclists may be associated with protesting against cycling infrastructure or greater support for "cycling licence" programs' (Delbosc et al. 2019, 688).

Their findings are consistent with those of Williams (2018) who analysed radio broadcasts in Ireland and found high percentages of dehumanisation in her corpus. She explains that cyclists are 'denied the human quality of being visible, denied the human instinct for self-preservation, and assigned behaviour that does not comply with societal norms and values'. These negative attitudes can have more serious consequences consisting in deliberately harming cyclists by creating dangers for them in the way drivers use their cars or by leaving tack or wires on bike paths. While, as Delbosc et al. point out, causation has not been conclusively established by these investigations, it is reasonable to hypothesise a correlation between negative attitudes and aggressive behaviour. This is why it is useful to take into account existing work concerning other forms of violence existing in society (against ethnic and religious minorities or against women for example) and envisage physical violence as the extreme manifestation of a continuum ranging from mild forms of discriminatory discourse to escalating examples of aggressive attitudes. In fact Delbosc et al. (2019, 686) asked their sample to report harassment (shouting and rude gestures) and direct aggression escalating from using one's car to deliberately block a cyclist or drive close to them to using one's car to deliberately cut off a cyclist or throw an object at them.

The aim of the present work is to observe these issues from the perspective of discourse. While discourse is certainly not the only perspective from which mobility choices are to be observed, and changes in discourse only are not likely to provide solutions to the problem, such changes could prove relevant in order to achieve long-term results. This book aims to bring light onto the role that language plays in the complex debate around mobility choices. In doing so, it will refrain from looking at language as separate from the society by which language is used, and will rather attempt to detect and follow threads within the complexity of voices burgeoning around this topic. Blommaert's use of complexity theory proved helpful as a source of inspiration for his approach to the study of sociolinguistics both in the ways it approaches language as a complex system and in how it analyses the ways in which semiotic activity—the use of signs—turn spaces into non-neutral and even agentive zones (Blommaert 2013, 15).

Blommaert's work proved significant for the investigation presented in this book not for its findings (the analysis concerns sociolinguistics and linguistic landscape) but for the inspiration he draws from chaos and complexity theory. He explicitly refers to the notion of entropy, the loss of energy that systems inevitably develop as a consequence of their state of non-equilibrium, and the natural tendency towards homogeneity and the reduction of diversity between two systems. As entropy is the central notion of the second law of thermodynamics, this brings us back to Rifkin's point (2011, 195) about how the laws of thermodynamics are better suited to be employed as metaphors in order to understand the world we inhabit. Blommaert significantly observes 'with regret' that even if these ideas are decades old and many of them have become common sense, they have not done so 'in sociolinguistics and many other branches of the human and social sciences, nor in public policy'. He adds: 'they have more influence and are much better understood in New Age movements than in the EU Commission or in any department of sociolinguistics, and that is a pity' (2013, 10).

It is with all these sources of inspiration, caveats and theoretical grounding that this book approaches the discourse of cycling. An observation made by Albert Einstein in a letter to his son, who was suffering from mental illness at that time, nicely wraps up the effectiveness of metaphors and the notions of energy and balance: '*Beim Menschen ist es wie beim Velo. Nur wenn er faehrt, kann er bequem die Balance halten.* (It is the same with people as it is with riding a bike. Only when moving, can one comfortably maintain one's balance)' (as quoted in Guroff 2016, 87).

We can thus start our ride.

References

Andrews, N., Clement, I., & Aldred, R. (2018). Invisible Cyclists? Disabled People and Cycle Planning—A Case Study of London. *Journal of Transport and Health, 8*, 146–156.

Angermuller, J. (2015). Discourse Studies. In J. D. Wright (Ed.), *International Encyclopedia of the Social & Behavioral Sciences* (2nd ed., Vol. 6, pp. 510–515). Oxford: Elsevier.

Angermuller, J. (2018). Truth After Post-truth: For a Strong Programme in Discourse Studies. *Palgrave Communications, 4*(30), 1–8.

Angermuller, J., Maingueneau, D., & Wodak, R. (Eds.). (2014). *The Discourse Studies Reader: Main Currents in Theory and Analysis*. Amsterdam: John Benjamins.

Basford, L., Reid, S., Lester, T., Thomson, J., & Tolmie, A. (2002) *Drivers' Perceptions of Cyclists. Prepared for Charging and Local Transport Division, Department for Transport* (TRL Report TRL549). https://trl.co.uk/sites/default/files/TRL549.pdf. Accessed 16 May 2019.

Blommaert, J. (2013) *Etnography, Superdiversity and Linguistic Landscapes: Chronicles of Complexity*. Bristol: Multilingual Matters.

Bloor and Bloor. (2007). *The Practice of Critical Discourse Analysis: An Introduction*. London: Routledge.

Delbosc, A., Nanzin, F., Haslam, N., & Haworth, N. (2019). Dehumanization of Cyclists Predicts Self-reported Aggressive Behaviour Toward Them: A Pilot Study. *Transportation Research Part F: Traffic Psychology and Behaviour, 62*, 681–689.

Fairclough, N. (1992). *Discourse and Social Change*. Cambridge: Polity Press.

Fairclough, N. (2016). A Dialectical-Relational Approach to Critical Discourse Analysis in Social Research. In R. Wodak & M. Mayer (Eds.), *Methods of Critical Discourse Studies* (3rd ed., pp. 86–108). London: Sage.

Fairclough, N., & Wodak, R. (1997). Critical Discourse Analysis. In T. van Dijk (Ed.), *Discourse as Social Interaction* (pp. 258–284). London: Sage.

Foucault, M. (1971). *L'Ordre du Discours*. Paris: Gallimard.

Guroff, M. (2016). *The Mechanical Horse: How the Bicycle Reshaped American Life*. Austin: University of Texas Press.

Halliday, M. A. K. (2001). New Ways of Meaning: The Challenge to Applied Linguistics. In A. Fill & P. Mühlhäusler (Eds.), *The Ecolinguistics Reader: Language, Ecology and Environment* (pp. 175–202). London and New York: Continuum.

Haslam, N., & Stratemeyer, M. (2016). Recent Research on Dehumanization. *Current Opinion in Psychology, 11*, 25–29.

Jäger, S., & Maier, F. (2016). Analysing Discourses and Dispositives: A Foucauldian Approach to Theory and Methodology. In R. Wodak & M. Meyer (Eds.), *Methods of Critical Discourse Studies* (3rd ed, pp. 109–136). London: Sage.

Johnson, M. (2014, September 17). Why Do We Still Hear People Joke About Hitting Cyclists? *The Conversation*. https://theconversation.com/why-do-we-still-hear-people-joke-about-hitting-cyclists-31214. Accessed 16 May 2019.

Kiernan, P. (2018). *Language, Identity and Cycling in the New Media Age*. London: Palgrave Macmillan.

Kubiszewski, I., Costanza, R., Franco, C., Lawn, P., Talberth, J., Jackson, T., & Aylmer, C. (2013). Beyond GDP: Measuring and Achieving Global Genuine Progress. *Ecological Economics, 93*, 57–68.

Lakoff, G. (2010). Why It Matters How We Frame the Environment. *Environmental Communication, 4*(1), 70–81.

Lakoff, G., & Johnson, M. (2003). *Metaphors We Live by*. Chicago and London: University of Chicago Press.

Lang, J. (2017). Explaining Genocide: Hannah Arendt and the Social-Scientific Concept of Dehumanization. In P. Baehr & P. Walsh (Eds.), *The Anthem Companion to Hannah Arendt* (pp. 175–196). London: Anthem Press.

Martin, J. R. (2004). Positive Discourse Analysis: Solidarity and Change. *Revista Canaria de Estudios Ingleses, 49*, 179–200.

Mautner, G. (2010). *Language and the Market Society: Critical Reflections on Discourse and Dominance*. London: Routledge.

Mulderrig, J. (2018). Multimodal Strategies of Emotional Governance: A Critical Analysis of 'Nudge' Tactics in Health Policy. *Critical Discourse Studies, 15*(1), 39–67.

Oosterhuis, H. (2019). Entrenched Habit or Fringe Mode: Comparing National Bicycle Policies, Cultures and Histories. In T. Myllyntaus & T. Männistö-Funk (Eds.), *Invisible Bicycle: Parallel Histories and Different Timelines*. Leiden: Brill.

Raser, E., Gaupp-Berghausen, M., Dons, E., Anaya-Boig, E., Avila-Palencia, I., Brand, C., … Panis, L. I. (2018). European Cyclists' Travel Behavior: Differences and Similarities Between Seven European (PASTA) Cities. *Journal of Transport & Health, 9*, 244–252.

Rifkin, J. (2011). *The Third Industrial Revolution*. New York: Palgrave Macmillan.

Sheller, M. (2018). *Mobility Justice: The Politics of Movement in an Age of Extremes*. London and New York: Verso Books.

Stibbe, A. (2018). Positive Discourse Analysis: Rethinking Human Ecological Relationships. In A. F. Fill & H. Penz (Eds.), *The Routledge Handbook of Ecolinguistics* (pp. 165–178). New York and London: Routledge.

Stiglitz, J. E. (2004). Evaluating Economic Change. *Daedalus Summer, 2004*, 18–25.

Stiglitz, J. E., Sen, A., & Fitoussi, J. P. (2010). *Mismeasuring Our Lives: Why GDP Doesn't Add Up*. New York: The New Press.

van Dijk, T. A. (2013). CDA Is Not a Method of Critical Discourse Analysis. *EDISO Debate—Asociacion de Estudios Sobre Discurso y Sociedad*. https://www.edisoportal.org/debate/115-cda-not-method-critical-discourse-analysis. Accessed 16 May 2019.

Williams, N. (2018, September) *Radio Discourse and Cycling*. Poster presentation at *Cycling and Society*, University of West England. http://www.cyclingandsociety.org/wp-content/uploads/2018/09/P_Williams.pdf. Accessed 16 May 2019.

Wodak, R., & Meyer, M. (Eds.). (2016). *Methods of Critical Discourse Studies* (3rd ed.). London: Sage.

CHAPTER 3

History, Urban Planning and Controversies

Abstract This chapter provides an overview of the major aspects linked to cycling advocacy or cycling policies, first from a historical perspective then focusing on contemporary cycling-related issues and findings. The role that non-linguists attribute to language and metaphors is highlighted, as in the metaphor of Darwinian evolution to frame the technological changes of velomobiles through history or the creation of 'jaywalkers'. Some history of urban planning is presented, focusing on Buchanan's work. The chapter then reports studies on 'bikelash', highlighting on discursive strategies to prevent it. At the end of the chapter some controversial issues that often arise when discussing cycling are briefly explained, these are mandatory helmet laws, road tax, safety in numbers and the exposure to pollution.

Keywords Velomobile · Jaywalking · Road tax · Bikelash · Safety in numbers

This chapter provides an overview of the major aspects linked to cycling advocacy or cycling policies, first from a historical perspective then focusing on contemporary cycling-related issues and findings. It does not report original findings but it brings together scientific data from an array of disciplines observing and blending them while highlighting the role that non-linguists attribute to language and metaphors. This chapter

M. C. Caimotto, *Discourses of Cycling, Road Users and Sustainability*, Postdisciplinary Studies in Discourse,
https://doi.org/10.1007/978-3-030-44026-8_3

is particularly aimed at readers who have no previous knowledge of issues concerning cycling policies and the history of everyday cycling. It also addresses some of the most widespread assumptions about cycling.

As reported by Horton et al. (2007, 8–9), academic studies on cycling traditionally belong to four disciplines, looking at the issue from an historical perspective, investigating the sociology of sport, analysing the medical effects and implications and, of course, everything that is related to engineering, design and planning—whether in terms of technical aspects concerning bicycles themselves or those concerning the public infrastructures that make cycling possible. More recently, contributions have also been added from the perspective of sociologists of science and technology, social and cultural geography and the politics of the bicycle and cycling. Horton et al. add that 'cycling is profoundly relevant to a whole range of important contemporary debates, about how we move around and with what consequences, about the appropriate pace and scale of everyday life, about how we treat our bodies, our communities and our planets, about the very viability of human futures' and this is why it is important that we reflect and discuss on cycling and society (2007, 10).

According to the World Cycling Alliance, promoting the growth of cycling represents a solution to many—11 out of 17—of the Sustainable Development Goals. Being affordable, it can contribute to the reduction of poverty (SDG1) and improve access to food markets, increasing nutrition options (SDG2). By increasing levels of physical activity and reducing pollution, it improves the well-being of people (SDG3). Improving cycling infrastructures helps achieve gender equality, as it provides women with access to water, schools, jobs that would not be accessible otherwise (SDG5). In fact, thanks to its role in allowing access to new destinations for women, cycling also played a central part in allowing the development of the feminist movement in the 1890s (Horton 2009, 2–3). Moreover, cycling is an energy-efficient form of transport (SDG7) and the cycling sector generates more jobs than any other transport sector for the same turnover (SDG8). Increasing the number of cyclists helps governments to create resilient infrastructure—independent from complex high-tech technology—and sustainable transport systems (SDG9). Being an efficient way of using scarce space in urban areas, it contributes to more inclusive, safe and sustainable human settlements (SDG11). Ensuring the transportation by bicycle of people and goods contributes to sustainable production and consumption patterns (SDG12). Integrating cycling into their climate action policies, governments at all levels can take immediate action to tackle climate change (SDG13). Finally, the participation of the cycling movement, civil society organisations and experts supports the global partnership for sustainable development (SDG17) (UNRIC 2018; ECF 2016).

3.1 History and Narratives

The development of the modern bicycle took place between 1870 and 1890, a time in which various factors that would allow the development of the bicycle industry came together: modern factories that would assemble components coming from multiple sources, the development of particular materials (steel, rubber), the promotion and marketing by distributors and advertisers (Smethurst 2015, 34). In the West, all this coincided with the rise of modernity and the bicycle was among the first industrial products that benefited from the techniques of advertising, marketing, distribution system and sponsorship, and in some cases, it pioneered them (35). It may appear ironic that an object, which nowadays is promoted as a sustainable, environment-friendly solution, started its history as an icon of modern consumerism. In fact, the part that the bicycle played in the development of consumerism is reflected in the negative attitude of left-wing twentieth-century writers who write about the cycling craze, criticising the status it gave to the activity of cycling (see Oddy 2007, 101).

The history of the bicycle is often framed as a series of technological changes which took place one after the other, starting from the most ancient machines that had some resemblance with the bicycle: the 1817 Draisienne or hobby horse (a machine without pedals that the user would move by pushing back with feet touching the ground), the 1867 French velocipede (with pedals and rotary crank fitted to the hub of the front wheel, which was slightly larger than the back wheel), the 1869–1870 Ordinary or 'Penny-farthing' (with a large front wheel with pedals, which the rider would drive from a seat above it and a much smaller back wheel) and the 1884–1886 safety bicycle, resembling the bicycle as we know it today, with a roller chain. (See Smethurst 2015, 16–21 for a detailed description with pictures of the various models.)

This historical approach focusing on technological evolution—Cox and Van De Walle (2007, 113) explain—invites us to see the bicycle as a technology that has been superseded by subsequent motorcycles and motor cars and, consequently, a machine that is fundamentally anachronistic in our contemporary society. In their work we see the effect that metaphors can have on our way of thinking: the authors convincingly demonstrate that the metaphor of evolution—derived from Darwin's theories—employed to observe the history of technology is likely to generate the conclusion that the bicycle's technology reached its acme in the 1890s and was then superseded by other new machines. These new machines are categorised as different and separate technologies, which evolved from the bicycle through radical change, for example with the addition of a motor to obtain the motorcycle (2007, 116).

Their thorough survey of technological innovations and of machines that do not fall within the dominant categories demonstrates that the narrative of progress that sees the automobile as the final product of a linear evolution—starting from the bicycle and passing through the motorcycle—is only one of the possible ways of looking at the available transportation options, and not the best one, given the actual fact that the three technologies actually coexist. The pseudo-Darwinist model that sees the automobile as the 'most evolved' form substituting the 'lesser' and inferior species, they explain, is deeply influential, as it generates a hierarchy in which any previous, 'obsolete' technology can only be envisaged as an archaic curiosity. They highlight that 'there is no inherent evolutionary process in technological development. The presence and persistence of particular technologies results from political, economic and social decisions' (119), 'Bicycles don't evolve, they are constructed' (129).

Cox and Van De Walle contrast this progression, that they label 'evolinear', with a 'non-evolinear organisation of personal transport' (2007, 126). They include the category of the velomobile, consisting in 'a form of pedal cycle (two, three or four wheeled) with a form of enclosure or bodywork serving both to protect the rider and to increase the aerodynamic efficiency of the vehicle' (114) and underline the importance of using the term 'velomobility' to signal the parallels and connections with research focusing on 'automobility' while distinguishing the specific concern for the materialities and practices of cycling (see Horton et al. 2007, 2).

Figure 3.1 (based on Cox and van De Walle 2007, 126) looks at the relationships between the various categories both horizontally—drawing a parallel between the morphological shift between the motorcycle and the automobile—and vertically—with a shift between the motorcycle

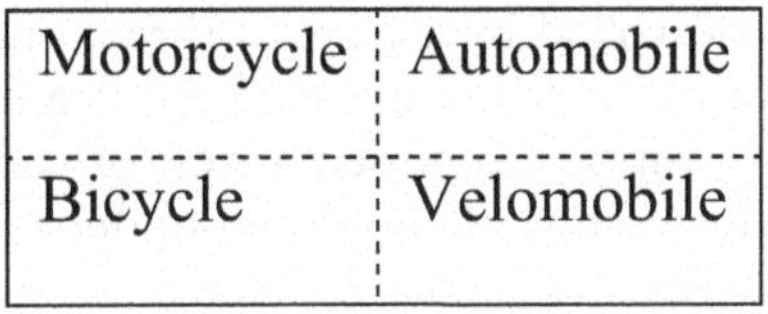

Fig. 3.1 Cox and van De Walle's diagram, with dotted lines

and the bicycle paralleled by the one between the automobile and the velomobile. The authors underline that the categories are not segregated from one another, which is why their diagram in here slightly modified by substituting the internal lines with dotted ones. The evolinear model embeds the assumption that the bicycle is a slow outdated vehicle, and this way of envisaging it is likely to result in infrastructures that consider cycling a slow, leisure activity. On the contrary, in urban environments, cycling allows people to move at an average speed higher than other forms of urban transport. Hence infrastructures that do not view bicycles—and other forms of velomobility—from this perspective are likely to prove unsatisfactory for everyday cycling.

Together with the contrast between various means of transport, the rivalry between people driving and people walking is also part of the development of the 'us vs. them' narratives we are investigating and it can help understand the current contrast between people who cycle and people who drive. Norton (2007) recounts the history of rivalry between drivers and pedestrians in the United States in the 1920s and 1930s. In his work discursive strategies are identified as a major tool employed to modify the perception and organisation of the streets and the introduction of traffic laws. Norton investigates the origin of the term 'jaywalking', explaining that 'jay' referred to a peasant with a provincial behaviour unfit for the more sophisticated city and extending it to their inability to walk in the city as required (2007, 342). When the presence of cars in cities was a novelty, many people resisted and claimed the public space of streets for walking and as a space where children were allowed to play. Norton reports many unsuccessful attempts to stop children and pedestrian from occupying streets (334–338) and an initial tendency to promote the enforcement of very stringent speed limits for cars that would have cancelled all the advantages that cars could offer. In fact, further back in 1865, in Britain the English Locomotives on Highways Act introduced permissible speeds on public roads to 3 km/h within cities and 6 km/h outside, and the requirement for each carriage to be controlled by a crew of three, one of which would have to walk in front of the vehicle carrying a red flag of warning (Reid 2015, 268).

Various representatives of automotive interests, or motordom—as they called themselves—understood that the social construction of the street was a major danger for the development of an urban future for cars, and the creation of the notion of jaywalker was a powerful discursive strategy to manufacture consent towards the legitimacy of cars and illegitimacy of other road users. Before taking its current meaning of 'someone crossing the street where not allowed or without paying attention to the traffic', jaywalker used to be employed by pedestrians to criticise other pedestrians who behaved in annoying ways, for example not moving aside to leave space to the person coming across in situations where they were expected to do so according to social norms. This, together with the reference to the countryside and a general implied superiority of city and progress against the country, helped to grant the success of the notion and the redefinition of streets as spaces 'where motorists belonged and where pedestrians were responsible for their own safety' (Norton 2007, 339).

By sponsoring road safety education in schools, motordom were able to create peer pressure and for example, Boy Scouts used to distribute cards to convince people they should not jaywalk (345). By 1925 the Los Angeles city council had introduced strict pedestrian control measures including fines for jaywalkers and Norton reports evidence that motordom aimed to influence the attitudes of newspapers (355) through their sponsoring of advertising. Attempts to create similar negative labels to refer to unwanted behaviour coming from drivers were less successful, 'jay riders' and 'joy riders'. As for the former, Norton highlights that 'A jay is foolish or gullible. Critics of motorists could call them cold-hearted, tyrannical, or selfish, but a motor car conferred too much power, modernity, and worldly sophistication to be the possession of a simpleton' (350). As far as 'joy rider' is concerned, it is hard to see how a metaphor pairing the idea of a car with the notion of 'joy' could effectively become a negative epithet to be used against drivers.

Norton's historical overview is helpful to understand attitudes towards people who cycle even if his work focuses on how the removal of people on foot from the streets came to be accepted socially. A particularly significant element is the shift of responsibility of collisions from people *inside* cars to the responsibility of people *outside* cars to think about their own safety (see also Baker 2019). Aldred (2013) has investigated the negative attitudes towards cyclists employing Goffman's concept of stigma, limited to the practices of othering, blaming and shaming (often called symbolic stigma). According to her findings, the cycling stigma impacts negatively on those who cycle, as there is a tendency to focus on 'bad cyclists' and to blame them for crashes, even when they are the ones who are injured. As Norton puts it 'Motorist's most persuasive claim to roadway priority [...] lay in the physical threat they represented to others' (2007, 350).

3.2 Urban Planning and Cycling Policies

Urban planning represents one of the most complex and challenging issues our contemporary society needs to tackle. Since 2009, according to the Population Division of the UN Department of Economic and Social Affairs, the world has been more urban than rural. In 2018, 55% of people lived in urban areas and this trend is expected to increase gradually in the upcoming years. This makes the provision of solutions for the sustainable development of cities and urban settlements one of the most important development challenges of our century, and the first three needs highlighted by the United Nations are housing, transportation and energy systems (UN DESA 2018). Moreover, according to the World Health Organization, increased urbanisation—with the consequent growth in violence, high-density traffic, low air quality and lack of parks and sports/recreation facilities—has contributed to the rising levels of physical inactivity, which accounts for 3.2 million deaths each year (WHO 2018).

From a historical perspective, one of the documents that has had a major impact in the way in which urban development has been envisaged is the British report *Traffic in Towns* (Buchanan 2015) which was originally published in 1963. It was a controversial document, that influenced traffic and town planning well beyond Britain, but its contradictions were not evident when it was first published and acclaimed. Its Penguin version received much publicity and sold 17,000 copies in the first four months. The aim of the report was to provide a response to the growing problem of traffic congestion, advancing possible solutions to adapt existing cities to the 'motor age' and the exponential growth of car ownership. While recent scholarship views modernism and conservationism as antithetical—Gunn explains (2011, 522)—in the Buchanan report they were yoked together, as the solutions proposed attempted to introduce strategies that would conserve the character of old town centres that were never designed for motor transport, while allowing people the freedom that was widely associated with car ownership, the idea of 'seamless' journeys.

In retrospect, the report introduced new ways of thinking about the 'environment'. It was explicitly defined as 'a convenient term [...] to convey the idea of a place, or an area, or even a street, which is free from the dangers and nuisances of motor traffic'. The choice of the term was described as 'clumsy' as it did not mean to convey the idea of an 'aesthetically stimulating' place, as most people would understand it; it simply meant to describe a place 'quiet, safe, clean, uncluttered by cars, and safe for children' (Buchanan 2015, 53). This quotation, with its repetition of the adjective 'safe' shows that safety, like in the 1920s and 1930s, was still perceived as the main problem. In fact, the new notion of 'environmental capacity' introduced by the report was related to the ability of people crossing the street without dangers, and was based on parameters of 'vulnerability' and 'protection' (2015, 219). One of the main criticisms levelled at the Buchanan report in subsequent years was that it ignored other forms of transport (Gunn 2011, 534) and took for granted that:

> the desire to own a car is both widespread and intense. The number of people who genuinely do not desire to possess their own private means of transport must be very small, and we think it is safe to base estimates of the future on the assumption that nearly all the families who, at any time, can afford to own a car (or who think they can) will in fact do so. (Buchanan 2015, 6)

Cycling in particular—even if the number of bicycles on the roads was still significant when the report was written—was silenced. The number of occurrences of the lemmas is 'car' 438, 'bicycle' 1. The only short passage referring to cyclists (286 words in total) states that 'it must be admitted that it is a moot point how many cyclists there will be in 2010', adding 'we have no doubt at all that cyclists should not be admitted to primary networks, for obvious reasons of safety and the free flow of vehicular traffic. It would make the design of these roads far too complicated to build 'cycle tracks' into them, nor would this be likely to provide routes convenient for cyclists in any case'. This and the next—and last—paragraph, worth quoting in its entirety, clearly show the car-dominant approach that was behind the whole report:

> In the conditions that are likely to arise in the future, as vehicular traffic grows, we are inclined to think that it will become necessary to divert cycle traffic increasingly to the less busy roads. In many instances it may even be desirable to allow cyclists and pedestrians to use the same route, suitably sub-divided, especially at critical points where cycle and pedestrian routes cross over or under main distributors. After all, as far as vulnerability to motor traffic is concerned, pedal cyclists and pedestrians have a great deal in common. (Buchanan 2015, 78–79)

We can observe here many aspects that would be considered particularly problematic by advocacy cycling groups: the focus on vulnerability—which generates fear of cycling and contributes to deterring people from using a bicycle to move around (Horton 2007), the idea that cyclists and pedestrians should share the same space—an approach that has proved to be problematic in many ways (Peck 2012), and of course the idea that vehicular traffic should be given priority over any other form of transport.

3.3 Environmental Politics and Bikelash

When comparing Buchanan's words to contemporary urban planning policies, we should refrain from a simplistic view envisaging a contrast between the individualistic and liberal car-dominated mentality promoted since the 1960s—and still exerting some influence nowadays—and the more recent approach focusing on making cities sustainable. It is useful to take a step back and observe the way in which the discourse of environmentalism has been incorporated into that of neoliberal policies, emphasising the link between sustainability and economic development. As Rossi and Vanolo (2011, 45–48) explain, in the field of sustainable urban planning 'resilience' has become the buzzword, presenting the environment not as a common good that should be protected from private interest but as an opportunity for economic gain. Or, in the words of Swyngedouw:

> There is indeed, in the domain of environmental policies and politics, a widespread consensus that Nature and The Environment need to be taken

> seriously, and that appropriate managerial-technological apparatuses can and should be negotiated to avoid imminent environmental catastrophe. At the same time, of course, there is hegemonic consensus that no alternative to liberal-global hegemony is possible. (2007, 28)

Lash and Urry (1994, 292–305) investigate the issue by elaborating on the implications of the environmentalist motto 'think globally, act locally' and exemplify the difficulties of implementing certain policies by referring to 'the need to reduce the number of car journeys in order to limit carbon emissions' (301) as a situation in which large numbers of people would need to act altruistically on behalf of temporally and/or geographically distant populations. They point out that 'people's commitment to and interests in their locality are varied and complex and this variety of interests will seriously affect the possibilities of large numbers of people "acting locally"' (304).

An example of local urban conflicting interests is the phenomenon described as 'bikelash', which consists in various forms of angry opposition within local communities when new cycling infrastructures are designed or built. According to Wild et al. (2017, 506) the focus of modern transport planning revolves excessively on a techno-centric and rationalist paradigm that does not allow it to engage with the sociopolitical dimensions of mobility. In particular, they explain that opposition to bike lanes is an expression of latent cultural and economic anxieties and underline the non-neutral and political nature of space for people who cycle: new cycle lanes inevitably disrupt a real or imagined order within the streetscape, a change that some will consider a new opportunity and others will perceive as a loss. Their analysis significantly engages with the fact that opposition is not monolithic, and in order to understand it, we need to identify and acknowledge the different reasons why people object to cycling in general or to specific cycling infrastructures.

Wild et al.'s investigation identifies four groups who are most likely to oppose the creation of new cycle lanes: retailers, anti-gentrification activists, marginalised cyclists, conservative voters (2017, 508). They describe the characteristics of these four views, providing possible solutions to reduce the level of conflict. Retailers oppose cycle lanes out of unfounded fears that a reduction in parking space will cause a reduction in their incomes, fears that usually dissipate a few months after the changes have been implemented. Most of the strategies proposed to help prevent or diminish their negative reactions revolve on discursive strategies:

promoting cycle lanes as improvements for the neighbourhood and the safety of its inhabitants rather than an improvement for cyclists only and involving retailers in collaborative design projects for the neighbourhood (2017, 509). The link between gentrification and cycle lanes is complex, while in most cases cycle lanes are a consequence of a gentrification process rather than a cause, Wild et al. underline the way in which 'urban revitalisation' processes are celebrated by politicians and cycling advocates without paying attention to the marginalisation of local inhabitants, who will perceive the improvement—which includes cycle lanes—that they had been asking for only catered when they are being pushed away (2017, 513).

Rejection of cycle lanes from people who cycle is often linked to poor planning, resulting in infrastructures that increase the danger of cycling rather than reducing it. In these cases, people who actually cycle are not consulted during the planning process and a way to prevent and reduce this form of bikelash, as in many other cases of (perceived) marginalisation, is extensive ongoing consultation. Negative reactions from people who cycle also belong to an older tradition of opposition against the segregation of cyclists perceived as a form of marginalisation, as it reinforces the idea that cyclists are not legitimate road users. In all these forms of opposition there are clear identity-related discursive strategies at work, generating a variety of 'us vs. them' narratives, made more complex by the fact that some of these appear in contrast among them, as in the case of anti-gentrification versus the role played by the remaining category, that of conservative voters. The latter is particularly interesting within the framework employed in this book and can be linked to the wider perspective of conservative politics refusing to implement changes to tackle the problem of climate change, either by negating climate change itself and its anthropogenic origin or by swallowing the ecological discourse within a wider neoliberal discourse that does not question the excessive exploitation of natural resources required to feed the dominant modern lifestyle (Lakoff 2010).

Wild et al. sum up the work of numerous authors who investigated the connections between conservative values and resistance to the creation of cycle lanes: 'the centrality of car travel to notions of family and economic responsibility; the commitment to suburbanism; a belief in market-led growth over state-led planning; and the associated commitment to privatised road space over the notion of streets as polis' (2017, 510). The attitudes reported, and the ways in which the contrast is envisaged, revolve around a perception of cyclists as 'others', portrayed as a 'threat' and 'outsiders' and, in some cases, also juxtaposing racist attitudes by stating or implying that most cyclists are 'illegal immigrants'. These attitudes require two responses from advocacy groups, one consisting in showing that people who cycle are 'regular people' and the other in promoting the fact that the presence of more people who cycle has positive effects in terms of safety for the whole population. From the discursive point of view, these othering strategies often share some characteristics that typically belong to discriminatory discourse. For example, Sternbergh (2011) has pointed out how anti-bike-lane arguments echo the Tea Party.

From a more distant perspective, Sheller and Urry (2000) reveal the pervasive role that cars have played in the creation of (post)modern lifestyles and what this entails for society and for democracy. They explain, 'if *urbanization* leads to the intensification of human habitats, the concentration of places in space, and the unification of condensed temporal flows, then *automobilization*, by contrast, leads to the extension of human habitats, the dispersal of places across space, the opportunities to escape certain locales and to form new socialities, and the fragmentation of temporal flows, especially through suburbanization' (2000, 742). In this sense, we see how automobility is dominant and profoundly rooted in contemporary societies, part of what Urry has described as a socio-technical system, in which social life became 'locked in' to the mode of mobility generated and presupposed by automobility (Urry 2004, 27).

Suburban housing, oil companies, out-of-town shopping centres all contribute to make the system hard to reverse and, Urry explains, we will never go back to a nineteenth-century public mobility dominated by buses, trains, coaches and ships, because of the need of individualised, fragmented and instantaneous mobility generated by the car system (2004, 36). But various changes will reach a tipping point during the twenty-first century and the car will be seen as a dinosaur, like fax machines, immobile phones or early freestanding PCs. A new 'post-car' system will be born, and 'there will be a mixed flow of slow-moving semi-public micro-cars, bikes, many hybrid vehicles, pedestrians and mass transport integrated into a mobility of physical and virtual access' (2004, 35). This perspective is useful for advocacy groups and policymakers: promoting cycling as part of a desirable society that responds to people's mobility needs better than the current car-dominated system could be a key to success.

The negative attitude towards cycle lanes coming from conservative voters and the negative attitude from people who cycle against segregation is linked to the notion of property vs. common good. The fierce opposition that can arise when discussing cycle lanes has to do with the idea that only certain road users are allowed to use certain parts of the road. Nowadays we take for granted the semiotic system that tells us where pedestrians are supposed to (not) walk and as children we learn that most parts of most streets are dangerous spaces where we should remain as little as possible and which are 'full of moving, dangerous iron cages' (Urry 2004, 30). In many countries, this semiotic system often ignores people who cycle and bicycles can thus be regarded as out of place wherever they travel. As we take the system for granted, we also tend to assume that the opposition between bicycles and cars has always been there. In fact, as reported by Reid (2015, 2) cyclists were the first ones who pushed to improve the quality of roads and create smooth surfaces. For example, the British Roads Improvement Association was created by cycling organisations in 1886. Many of these cyclists after became motorists and continued the campaigning they had started years before.

3.4 Controversial Issues

The opposition between cyclists and motorists is also behind many of the arguments that tend to resurface when cycling policies are being discussed. Car owners pay taxes and insurance, need a licence, must wear seat belts and have to be identifiable through plates. These facts generate the feeling in some car owners that they are more entitled than others to use roads and to use public space to store their private cars, envisaging parking as an absolute right and rejecting limitations to the free circulation of cars—such as congestion charges to limit access to restricted areas in cities. The so-called 'road tax' in particular is one of the arguments employed by drivers to justify their privileged ownership of roads, but 'road tax' was abolished in 1937 and substituted by Vehicle Excise Duty, which can be classified as pollution tax as it is based on the size of engine and emissions (Harrabin 2013). The expression 'road tax' suggests that roads are paid for by motorists, while in fact roads are paid for via general and local taxation, hence drivers, cyclists and pedestrians—in one word, citizens—all equally contribute to road maintenance. As far as parking is concerned, taking up public space with one's vehicle is actually an exception to the limitations and fees that most countries apply to other forms of occupation of public space. Tokyo—a place where land is scarce—notably restricts the occupation of streets with cars, as car buyers must provide proof to the police of parking space in order to register their car (Kanagawa Prefecture Police 2019). Because of this perception of privileged access, more limits and regulations for cyclists are often invoked, such as the need for a licence and a plate or mandatory helmet laws.

According to the UK Department for Transport, 80% of people who cycle also hold a driving licence (DfT 2012), hence the idea that cyclists do not behave properly on the road because they ignore traffic laws is flawed. In this respect, through the book we shall observe the effects of narrative as cyclists are generally expected to behave better than other road users and tend to be perceived as a more cohesive group when compared to drivers or pedestrians (see Furness 2010). A very common point is that cyclists 'always' ride through red lights, but according to Johnson et al. (2011, 323) red light non-compliance is not high and the same is true for cyclist crash involvement as a result of riding through a red light. Richardson and Caulfield (2015, 71) investigated red light running reality and perceptions in Dublin City and observed that the public's perception is far worse than the actual situation.

This negative attitude towards people who cycle is linked to the stigmatisation and dehumanisation mentioned in Chapter 2, cyclists are denied the human characteristic of self-preservation and are thus perceived as stubbornly rejecting the notion of road users' responsibility that was introduced and promoted by motordom, as explained by Norton (2007). The idea that responsibility on the road is linked to one's vulnerability, rather than to the level of threat one's vehicle and potential speed pose to other road users, is reflected, in cycling-related discourse, in the prominence of policy discussions to make some safety-improving features mandatory for cyclists, such as wearing a helmet or high-visibility clothes. Mandatory helmet laws have been demonstrated to have a large unintended negative impact in terms of net societal health benefits, as they cause a reduction in cycling and a consequent reduction in exercise and, most likely, a rise in car use, resulting in a less healthy environment for the whole population, while their impact in terms of reducing head injury rates is not sufficiently significant to justify the choice of introducing mandatory laws (De Jong 2012).

Helmets are not designed to resist impact with motorised transport and thus they do not protect from the most dangerous kind of collision that could occur. Data from the UK Department for Transport show that cyclists and pedestrians have similar levels of fatality risk (DfT 2013, 178), hence if cycling helmets are made mandatory, pedestrian helmets should be introduced as well. The presence of mandatory helmet laws is also likely to hinder the growth of bike-sharing schemes, as the need of carrying one's helmet along significantly reduces the level of spontaneity of choosing to cycle that represents one of the most relevant reasons for choosing a bike-sharing scheme over using a private bicycle. On average, when mandatory helmet legislations are brought in, there is a reduction of up to 30% in cycling and according to the European Cyclists' Federation 'it is [...] morally wrong and legally unjustifiable to pick out cycling as a high risk activity that merits mandatory helmet legislation, or for public authorities to campaign for their use' (ECF 2014).

A reduction in cycling also affects the Safety in Numbers effect, increasing the risk of injury to remaining cyclists. The Safety in Numbers effect describes the frequency of collisions in relation to the number of people walking and cycling. Studies have demonstrated that on average a modal shift from car to bicycle generates health benefits that are substantially larger than the risks related to cycling; this positive effect on individuals is even larger for society as a whole, due to the reduction of emissions and the lower number of traffic accidents (Johan de Hartog et al. 2010; Gössling et al. 2019). Safety in Numbers investigates specifically whether and how much the number of collisions between motorists and people walking or cycling increases or decreases when the number of people walking or cycling increases. Several studies have demonstrated that the relationship is not linear, and a rise in the number of people walking and cycling results in a higher decrease in the individual risk of being hit by a car. On the contrary, an increase in the number of cars generates a linear increase in the number of collisions (see Jacobsen et al. 2003). The implications of these findings are that it is important to understand the Safety in Numbers effect in order to introduce policies to promote cycling and walking without overestimating the risks—as such overestimation would hinder a modal change that can positively affect populations.

What causes the Safety in Numbers effect has only been partially demonstrated. An important aspect is that people who drive are required to perform a number of complex cognitive tasks that is higher than what the brain is able to handle, hence unexpected and/or smaller objects are less likely to be perceived. The possible policy responses to this are to decrease speed—thus allowing more time to process information—and increase the number of bicycles and people walking, thus making them less unexpected (Jacobsen et al. 2003). Jacobsen et al. (208) also refer to the legitimisation of cycling as a possible result of an increase in the number of people who cycle which could also contribute to explain the Safety in Numbers effect, but they state that more research is required in this respect.

Among the various health benefits brought by an increase in cycling, that of breathing better air is scarcely known. Common wisdom expects cyclists to inhale the worst amount of exhaust fumes, while on the contrary several studies demonstrate that 'car drivers are consistently exposed to the highest average levels of carbon monoxide and ultrafine particles' (Kingham et al. 2013). People who cycle spend less time in traffic, as they move more swiftly and tend to take smaller roads away from the worst congested areas. As a consequence their exposure to particulate is lower than that of people inside a car—car's air intake brings the exhaust fumes of other cars inside the car itself—but also of people on buses and people walking.

3.5 Concluding Remarks

When taking part in cycling advocacy activity (in many cases a reaction to the feeling of powerlessness in front of the arrogance and consequent danger of some people driving) it is tempting to divide the world into good-environmentalist-smarter cyclists and bad-selfish-dumb-drivers who are protected and sustained by powerful-rich-all-controlling-car industries while the majority of people have accepted and will never question the dominance of cars in our cities and in our lives (except cyclist heroes of course). It is a self-satisfying narrative, but one which, as any other 'us vs. them' narrative, is flawed from the start. Investigations into the history of cycling, car-manufacturing and urban planning reveal the complexity and the shades we cycling lovers often ignore or prefer to ignore, as the common history of the first cyclists and motorists (Reid 2015) or the role of the bicycle in the birth of consumerism and how elitist cycling was at the very beginning. Only if we dig into these histories, we revisit the pseudo-Darwinian evolution of bicycles, we admit that as human beings alive in present days we are part and parcel of the market society and we all perceive that we have mobility needs in terms of speed, reliability and independence that were in fact generated by automobilisation (Sheller and Urry 2000, 742), only if we take into account all these less obvious aspects we can take a step back, realise what we have in common with our reified 'motorist enemies' and start afresh.

Those of us who have never taken part in advocacy activity, maybe never used a bicycle as a means of transport or don't even have one rusting away in their cellar can learn about a clash for power that is taking place around them and that entails a fascinating number of social issues and rights that concern all of us, even those who do not cycle and never intend to start. Battles to promote more cycling entail within an apparently small issue—one that policymakers may underestimate and expect to be solved by simply painting a few more cycle lanes—a vast number of implications, layers and assumptions that reflect the difficulties met on a much wider scale when communicating climate change and framing it so that people will understand its anthropogenic origins and react accordingly. When we revisit the history of mobilities, when we observe the rivalry between people gradually reified into fighting tribes, when we realise how terms that we take for granted actually carry hidden political and social implications, when we question our assumptions about how we share public space, only then we start realising the complexity and relevance of it all. In the next four chapters, we will observe real examples of policies, media reports and speeches from interviews in order to bring to the surface the discursive practices at work.

References

Aldred, R. (2013). Incompetent or Too Competent? Negotiating Everyday Cycling Identities in a Motor Dominated Society. *Mobilities, 8*(2), 252–271.

Baker, P. C. (2019, October 3). Collision Course: Why Are Cars Killing More and More Pedestrians? *The Guardian*. https://www.theguardian.com/technology/2019/oct/03/collision-course-pedestrian-deaths-rising-driverless-cars. Accessed 23 November 2019.

Buchanan, C. (2015 [1963]). *Traffic in Towns—A Study of the Long Term Problem of Traffic in Urban Areas. Introduction by Simon Gunn*. London and New York: Routledge.

Cox, P., & Van De Walle, F. (2007). Bicycles Don't Evolve: Velomobiles and the Modelling of Transport Technologies. In D. Horton, P. Rosen, & P. Cox (Eds.), *Cycling and Society* (pp. 113–132). London: Routledge.

De Jong, P. (2012). The Health Impact of Mandatory Bicycle Helmet Laws. *Risk Analysis, 32*(5), 782–790.

DfT. (2012). *THINK! Cyclist.* https://www.gov.uk/government/publications/think-cyclist. Accessed 10 June 2019.

DfT. (2013). *Reported Road Casualties Great Britain 2013.* https://www.gov.uk/government/uploads/system/uploads/attachment_data/file/359311/rrcgb-2013.pdf. Accessed 20 May 2019.

ECF. (2014). *ECF Helmet Factsheet.* https://ecf.com/files/wp-content/uploads/Helmet-factsheet-_17042015_Final.pdf. Accessed 10 June 2019.

ECF. (2016). *Cycling Delivers on the Global Goals.* https://ecf.com/groups/cycling-delivers-global-goals. Accessed 20 May 2019.

Furness, Z. (2010). *One Less Car: Bicycling and the Politics of Automobility.* Philadelphia: Temple University Press.

Gössling, S., Choi, A., Dekker, K., & Metzler, D. (2019, April). The Social Cost of Automobility, Cycling and Walking in the European Union. *Ecological Economics, 158,* 65–74.

Gunn, S. (2011). The Buchanan Report, Environment and the Problem of Traffic in 1960s Britain. *Twentieth Century British History, 22*(4), 521–542.

Harrabin, R. (2013). *Is There Any Such Thing as 'Road Tax'?* BBC News Website. https://www.bbc.com/news/magazine-23694438. Accessed 10 June 2019.

Horton, D. (2007). Fear of Cycling. In D. Horton, P. Rosen, & P. Cox (Eds.), *Cycling and Society* (pp. 133–152). London: Routledge.

Horton, D. (2009). *Social Movements and the Bicycle.* https://thinkingaboutcycling.files.wordpress.com/2009/11/social-movements-and-the-bicycle.pdf. Accessed 4 June 2019.

Horton, D., Rosen, P., & Cox, P. (2007). *Cycling and Society.* London and New York: Routledge.

Jacobsen, P. L. (2003). Safety in Numbers: More Walkers and Bicyclists, Safer Walking and Bicycling. *Injury Prevention, 9,* 205–209.

Johan de Hartog, J., Boogaard, H., Nijland, H., & Hoek, G. (2010). Do the Health Benefits of Cycling Outweigh the Risks? *Environmental Health Perspectives, 118*(8), 1109–1116.

Johnson, M., Newstead, S., Charlton, J., & Oxley, J. (2011). Riding Through Red Lights: The Rate, Characteristics and Risk Factors of Non-compliant Urban Commuter Cyclists. *Accident Analysis and Prevention, 43*(1), 323–328.

Kanagawa Prefectural Police Department. (2019). *Application Procedure for a Vehicle Parking Place (Garage) Certificate.* https://www.police.pref.kanagawa.jp/eng/e_mes/engf4001.htm. Accessed 20 June 2019.

Kingham, S., Longley, I., Salmond, J., Pattinson, W., & Shrestha, K. (2013). Variations in Exposure to Traffic Pollution While Travelling by Different Modes in a Low Density, Less Congested City. *Environmental Pollution, 181,* 211–218.

Lakoff, G. (2010). Why It Matters How We Frame the Environment. *Environmental Communication, 4*(1), 70–81.

Lash, S., & Urry, J. (1994). *Economies of Signs and Space*. London: Sage.

Norton, P. D. (2007). Street Rivals: Jaywalking and the Invention of the Motor Age Street. *Technology and Culture, 48*(2), 331–359.

Oddy, N. (2007). The Flaneur on Wheels? In D. Horton, P. Rosen, & P. Cox (Eds.), *Cycling and Society* (pp. 97–112). London: Routledge.

Peck, C. (2012). *New Guidance on Shared Use Routes*. https://www.cyclinguk.org/article/campaign-article/new-guidance-on-shared-use-routes. Accessed 20 May 2019.

Reid, C. (2015). *Roads Were Not Built for Cars—How Cyclists Were the First to Push for Good Roads and Became the Pioneers of Motoring*. Washington, Covelo, and London: Island Press.

Richardson, M., & Caulfield, B. (2015). Investigating Traffic Light Violations by Cyclists in Dublin City Centre. *Accident Analysis and Prevention, 84*, 65–73.

Rossi, U., & Vanolo, A. (2011). *Urban Political Geographies: A Global Perspective*. London: Sage.

Sheller, M., & Urry, J. (2000). The City and the Car. *International Journal of Urban and Regional Research, 24*(4), 737–757.

Smethurst, P. (2015). *The Bicycle—Towards a Global History*. London: Palgrave Macmillan.

Sternbergh, A. (2011). *'I Was a Teenage Cyclist,' or How Anti-Bike-Lane Arguments Echo the Tea Party*. https://6thfloor.blogs.nytimes.com/2011/03/09/i-was-a-teenage-cyclist-or-how-anti-bike-lane-arguments-echo-the-tea-party/. Accessed 10 June 2019.

Swyngedouw, E. (2007). Impossible "Sustainability" and the Postpolitical Condition. In R. Krueger & D. Gibbs (Eds.), *The Sustainable Development Paradox: Urban Political Economy in the United States and Europe*. New York and London: The Guilford Press.

UN DESA. (2018). *World Urbanization Prospects 2018*. https://population.un.org/wup/. Accessed 20 May 2019.

UNRIC. (2018). *How Does Cycling Help Achieve the Global Goals?* https://www.unric.org/en/latest-un-buzz/31011-how-does-cycling-help-achieve-the-global-goals. Accessed 20 May 2019.

Urry, J. (2004). The 'System' of Automobility in Theory. *Culture & Society, 21*(4–5), 25–39.

WHO. (2018). *Physical Inactivity: A Global Public Health Problem*. https://www.who.int/dietphysicalactivity/factsheet_inactivity/en/. Accessed 20 May 2019.

Wild, K., Woodward, A., Field, A., & Macmillan, A. (2017). Beyond 'Bikelash': Engaging with Community Opposition to Cycle Lanes. *Mobilities, 13*(4), 505–519.

CHAPTER 4

Stigmatisation in Newspapers

Abstract This chapter investigates how cyclists are portrayed in the media by exploring discursive moves that generate an 'us vs. them' narrative when cyclists and pro-cycling activists are discussed. It focuses, in particular, on the reporting of a fatal collision in London between Alliston, who was riding a bicycle, and Briggs, who was crossing the road. The qualitative analysis presented here observes a selection of newspaper articles concerning that collision, and the ensuing trial, by applying the approach proposed by van Dijk (Discourse and Society 3:87–118, 1992) to investigate racist discourse in the press. The discourse moves employed by journalists to deny racism, for face-saving reasons and identified by van Dijk in his research, are employed here to analyse the selected articles in order to show what racist discourse and the discursive stigmatisation of cyclists have in common. The chapter closes with a description of Poli's (*Mobility and Environment: Humanists Versus Engineers in Urban Policy and Professional Education*. Springer, Heidelberg, 2011) three approaches to traffic policy.

Keywords Collision · Racist discourse · Face-saving · Stigma · Us vs. them

M. C. Caimotto, *Discourses of Cycling, Road Users and Sustainability*, Postdisciplinary Studies in Discourse,
https://doi.org/10.1007/978-3-030-44026-8_4

The widespread perception of people who cycle is one of the main issues that need to be addressed in order to improve their levels of safety and to increase the number of people who choose a bicycle as a means of transport. At the time of writing, the most recent survey observing how cyclists are portrayed in the media focuses on online news in Scotland and the rest of the UK between March 2017 and March 2018. The main findings are that 61% of articles had a broadly negative sentiment, 'criminal act' and 'safety' being prominent themes that are not likely to encourage readers to take up cycling (Sustrans 2019, 3). Consistently with the previous findings of Rissel et al. (2010), the study shows that 'cycling' is portrayed more positively and 'cyclist(s)' more negatively. Moreover, middle-aged white men are over-represented, reinforcing gendered stereotypes (Sustrans 2019, 10). Aldred (2016) explains that there is relatively little literature on the relationship between the media and public understanding of transport. Her analysis of the relationship between cycling policy and media in London comes to positive conclusions, reporting an improvement in the way cycling is discussed in the media as a consequence of the changes in policy. Ralph et al. (2019) have observed the local news coverage of road collisions involving Vulnerable Road Users (VRUs), highlighting that 'subtly but consistently' VRUs are blamed while in most cases the drivers have no agency, as it is the 'vehicles' that hurt, collide and kill, rather than the people who should be in control of them. Goddard et al. (2019) show how linguistic strategies employed in collision news coverage influence readers' interpretation of what happened and what changes in discourse could be implemented.

This chapter aims to contribute to the scarce amount of studies that have investigated how cyclists are portrayed in the media by exploring the kind of discursive moves that generate an 'us vs. them' narrative when cyclists and pro-cycling activists are discussed. Walker sums up the issue in a video in which he discusses how cyclists are perceived and judged: 'In 2016, 448 pedestrians were killed on UK roads. Of these pedestrian deaths, just two involved cyclists. But one of the cases attracted so much media attention, the government quickly pushed through a new law making it easier to prosecute cyclists' (*The Guardian* 2019). The case he is referring to is the collision that occurred in London between Charlie Alliston, who was riding a bicycle, and Kim Briggs, who was crossing the road, in February 2016. Briggs died of brain damage one week later.

The qualitative analysis presented here observes a selection of newspaper articles concerning that collision, and the ensuing trial, by applying the approach proposed by van Dijk (1992) to investigate racist discourse in the press. The discourse moves employed by journalists to deny racism, for face-saving reasons, and identified by van Dijk in his research, are employed here to analyse the selected articles in order to show what racist discourse and the discursive stigmatisation of cyclists have in common. Aldred (2013) employs the notion of stigma to investigate how people who cycle perceive themselves and highlights the complexity of this self-image, which presents some differences if compared to other stigmatised groups. She argues that two stigmatised images of 'the cyclist' coexist, as a cyclist can be cast either as incompetent or as too competent. Results from her research show that cycling is still perceived as 'not normal' in societies that are 'socially and spatially dominated by motor vehicles to the detriment of other road users' (2013, 253).

While bad driving is seen as an individual attitude, bad cycling tends to be attributed to cyclists in general (Aldred 2013, 254; Furness 2010, 131–133). This stigmatisation triggers forms of discrimination and threat that can be assimilated to those observed in racist discourse. In the worst cases, this kind of attitude has led to the deliberate attack and killing of cyclists either because they were manifesting for their right to safety on the road or because they had an animate discussion with someone driving a car (Sparks 2011; Spagnolo 2016). These forms of extreme violence are often presented in the news as acts carried out by some lunatics, while in fact they can be envisaged as the extreme point of a continuum that links them to the everyday, normalised forms of discrimination, which are generally accepted or at least tolerated by the majority of the population (Johnson 2014). Of course this is true for many other forms of everyday discrimination—based on sex, religion or skin colour—that in some cases are pushed to the extreme of killing a member of the discriminated group for the simple reason of him/her belonging to that group.

As everyday discrimination is perpetrated, reinforced and normalised mainly through discursive means, a CDS approach helps to identify the elements that contribute to maintaining the idea that 'cyclists are not normal' (Sustrans 2019, 8) and that allow for one single death to become more prominent and perceived as more dangerous and outrageous than the remaining 446 pedestrian deaths and the fact that the main contributory factor reported in all reported casualties was 'Driver/Rider careless, reckless or in a hurry' (DfT 2017). The collision reported in the articles selected for this analysis was a prominent news story, for the first time in Britain a cyclist was accused of manslaughter and the judge found him not guilty. Alliston was condemned to two years for 'wanton and furious driving', an old law from the nineteenth century, because his fixed gear bike was not legal to ride on the road. Rather than focusing on the details of the collision and the court judgments, the goal of this investigation is to observe how the topic of Alliston's court case was employed by a variety of journalists to discuss cycling-related issues in general and—in many cases—(re)produce a very negative mental model of cyclists and cycling activists (see also Sustrans 2019, 51). Fincham (2007) shows examples of how the negative image of couriers portrayed in the media has influenced the negative representation and perception of cyclists in general. When reporting information about Alliston, newspapers often focused on his declared past work experience as a courier in London. For example, *The Guardian* announced his conviction with the headline 'Ex-courier convicted for mowing down woman on his track bike' (Grierson 2017).

The articles selected for analysis are from three British mainstream newspapers (*The Times*, *The Guardian*, *The Independent*) and one tabloid, *The Scottish Daily Mail*. The aim of this analysis is not to imply this kind of discourse is representative of the British press in general. This consists in a qualitative, fine-grained analysis to detect specific discourse moves that the Alliston's case allowed to come to the fore more clearly and openly than usual, thus providing space in mainstream media to express what would more usually be heard in informal conversations. Before moving to the actual analysis of the news articles, a summary of van Dijk's work on the discursive strategies employed to deny racism is provided.

4.1 Discourse and the Denial of Racism

As explained by van Dijk (1992), negations of racism are more likely to be found in societies in which discrimination and racism are officially banned, and any discourse that may be perceived as racist is likely to be condemned. This generates a strong necessity for individuals and institutions to prevent accusations of racism through face-saving strategies and discursive moves that pre-empt criticism. These face-keeping strategies are enacted through a series of six discursive moves that van Dijk labels 'positive self-presentation, denial and counter-attack, moral blackmail, subtle denials, mitigation, defence and offence' (1992, 103–107).

The parallel between discourse against cyclists and racist discourse was noticed by readers of *The Independent*. The headline of the article about the court case was 'Cyclists are rude, abusive and uncivil – but they do deserve more protection on our roads' (Street-Porter 2017) and one of the readers' comments read 'cyclists are the new Muslims. One Charlie Alliston and they're all Charlie Allistons' thus drawing a parallel with the discriminatory attitude experienced by all Muslims after some so-called Muslims had carried out terrorist attacks. Another comment was 'what a stupid headline. Cross out "cyclists" and then put in the "j" "n" or "p" words and read out aloud. How does it sound now. You should be ashamed of yourself', which was met with a further reply from another reader 'The difference is that cyclists are made, or rather, make themselves; whereas Js, Ns, and Ps were born like that'. This very last comment, with its profoundly racist implication that minorities should suffer in silence but cyclists are even worse because they actively search for discrimination, is revealing of the way in which a cyclist's identity is a reification.

The connection between the stigmatisation of cyclists and racism goes beyond discursive moves. As reported by Wild et al. (2018, 509–510), the characterisation of cyclists as 'outsiders' often has a racist element and there are cases in which the proposal of new cycle lanes was met with criticism because the lanes would favour terrorism, they would be employed by 'illegals' and would make the city 'look like Beijing'. At the same time, as already mentioned in Chapter 3, they point out the complex relationship between the creation of cycle lanes and gentrification, which often entails the racist action of evicting low-income communities of colour from their homes (Wild et al. 2018, 513). Nevertheless, this aspect is beyond our goals, and the connection here is simply employed to enrich the discursive analysis without focusing on the actual link between cycling and racism.

Coming back to the six moves identified by van Dijk, we shall briefly illustrate how they work in racist discourse and then we shall apply them to our case study. *Positive self-presentation* consists in an argumentative denial of the accusations of the anti-racists, reinforcing the idea that the journalist belongs to a tolerant group and that racism does not exist in their society, while the problems discussed in the text are ascribable to some reasons that have nothing to do with racism. *Denial and counter-attack* take the positive self-presentation further and van Dijk identifies three moves within this one: (1) the presupposition of 'truth'—what the journalist is saying is presented as an irrefutable fact (2) the consequent move is that of reversal '*we* are not the racist, *they* are the "true racists"' (3) the following implication is that those accused of racism are 'the real victims' and the anti-racists are 'the real enemy', hence these 'real victims' become 'the heroes' who 'dare' the 'anti-racist brigade' (1992, 104). The next move identified in the denial of racism is that of *moral blackmail*, which consists in the pretence of censorship: the journalists depict themselves as brave enough to break the taboo and state the truth even if they are morally blackmailed. Of course, van Dijk explains, the taboo is only imaginary and their articles actually reach a vast number of readers.

While the above-mentioned moves are relatively easy to recognise, the following ones are less evident. *Subtle denials* consist in moves to express doubt or distance through linguistic devices such as the use of inverted commas and of quoting verbs like 'claim' or 'allege'. *Mitigation* is often expressed through concessions that recognise some partial truth in what the 'enemy' states, another move that contributes to the construction of a positive self-presentation. Most of the time, mitigation is the move employed to introduce some further attacks, thus rendering them more acceptable. The final move, *defence and offence*, consists in open attacks against the anti-racists, often described as 'ideologically blinkered'. In the next section, we shall closely analyse a selection of articles concerning the court case of Charlie Alliston to show how this controversial topic was employed by some journalist as the gateway to allow a treatment of cyclists in general, and of cycling advocates, which was more negative than what would usually be deemed acceptable in mainstream media.

4.2 Headlines

The four articles selected for this analysis were retrieved online on the websites of the respective newspapers. Before moving to the application of van Dijk's categories, it is useful to present and closely observe the headlines. The one in *The Independent*, which we already mentioned, employs the typical structure with a premise presented as an irrefutable fact followed by the negative conjunction 'but' and some other opinion presented as fact: 'Cyclists are rude, abusive and uncivil – but they do deserve more protection on our roads' (Street-Porter 2017). If we compare this to the traditional 'I am not racist, but…' followed by a negative opinion of the speaker about blacks presented as a fact, we see that Street-Porter's headline is reversed, as the negative opinion is fronted while the second part of the sentence states something apparently positive.

The headline lists the three negative characteristics that are associated with cyclists by attributing them to the whole community and employs the existential (Halliday 1994) verb to be, implying this is not an opinion but a self-obvious fact. To apparently mitigate and balance this first half of the headline, the second part of the headline is introduced by a negative conjunction, which would supposedly be followed by something positive about cyclists. The next move creates an 'us vs them' dichotomy as 'they' deserve more protection on 'our' roads. This apparent concession to the cyclists' community in fact reinforces their dehumanisation as they are treated as a group that 'deserves' protection, thus removing all agency from them. At the same time, the nominalisation of 'protection' removes the agency of who should enact this protection, while the possessive adjective 'our' before roads is conveniently unclear in its referent, as it could refer to the fact that roads belong to everyone, or to 'our national roads'. Still, the semantic construction generates a contrast between 'they' (i.e. cyclists) and 'our roads' and when reading it is natural to infer that 'our roads' belong to someone different than cyclists—which corresponds to the dominant view in our car-centred society (Urry 2004).

Some similar implication is found in the headline and in the article published in *The Guardian* by Shariatmadari (2017) 'After Kim Briggs's death, cyclists must realise that they are traffic too'. Here, both the headline and the article have a more positive attitude towards cyclists. Still, as our close-grained analysis aims to demonstrate, in this moderate approach we can see the force of hegemonic discourse. The author tries to strike a balance between dialoguing with cyclists, rather than simply accusing them, but also writing something that non-cyclists will deem appropriate in the very negative climate generated by the collision and ensuing death. In doing so—perhaps unconsciously—he inserts in his articles subtle implications that reinforce and confirm a stigmatisation of cyclists.

The headline shows a prescriptive attitude, through the use of the modal verb 'must', and distancing, through the use of the pronoun 'they', even if the journalist declares to be a cyclist. The word 'traffic' simply refers to 'the passing to and fro of persons, or of vehicles or vessels, along a road, railway, canal, or other route of transport; also with reference to air travel' (OED 2019), but searching for the word in the British National Corpus shows that the lemma carries a negative semantic prosody (Kilgarriff et al. 2014). Thus stating that 'cyclists are traffic' infers that cyclists (again, reified as a group) consider themselves too positively and are not able to realise the consequences of their actions. As we shall observe later, the article attempts to convey a positive message but it is exactly in its subtle discursive implications that it reinforces stigmatisation, possibly against the intentions of the journalist himself.

Given the subtlety of the discriminatory discursive moves in *The Guardian's* article, especially when compared to the other articles scrutinised here, in order to fully recognise their presence and effectiveness, it can be helpful to substitute the stigmatised group (cyclists) with the dominant group (drivers)—a strategy often employed to make subtle sexism or racism more evident. In this case, the headline states something obvious on the literal level: cyclists are traffic, i.e. cyclists are part of the people moving with vehicles along roads. Nevertheless, the headline implies, they refuse to recognise this obvious fact. Now, because Kim Briggs was killed by someone riding a bicycle, they must recognise they are traffic too (which, by the way, also generates the implication that an intrinsic and normal characteristic of traffic is to kill pedestrians).

If we imagine to substitute the name of Kim Briggs with the name of one of the thousands of pedestrians killed by a collision with a car every year, followed by 'drivers must realise they are traffic too'—or even its more direct meaning 'drivers must realise they can kill too'—we can see this would be perceived as ludicrous, as the dominant perception is that bad driving is an individual attitude (Aldred 2013, 254), and that drivers as a group already realise they can kill because they have to obtain a driving licence in order to drive. This idea is in contrast with the available data: many drivers kill because they underestimate the potential tragic consequences of speeding, checking phones while driving, drinking and driving, and so forth (WHO 2018). Messages conveyed by public service advertisings—warning people of the potential consequences of speeding, texting while driving, driving when drunk, driving without wearing a seatbelt and the like (see DfT 2019)—also imply that not all drivers realise the potential consequences of their actions. Still, in the dominant discourse, a headline addressed at drivers like the one addressing cyclists in *The Guardian* would not be considered congruous.

The next headline we shall focus on is from *The Times* and is less transparent than the previous ones: 'Braking Good'. Readers are most probably expected to recognise an intertextual reference to the popular TV series *Breaking Bad* (Gilligan 2008) but there may be hidden implications that go beyond the phonetic similarity between 'braking' and 'breaking'. The title of the series is itself complex and has often been discussed by its fandom (Rothman 2013). According to the *Dictionary of American Regionalisms* (Hendrickson 2000, 30) the phrase is Southern American slang that means 'to behave in a violent manner for no good reason' which is what the journalist who authored the articles appears to think of Charlie Alliston's behaviour. It is possible and plausible to see a connection between Charlie Alliston and the TV series co-protagonist Jesse Pinkman: they are both young men, they have similar looks, they dropped out of school and do not have a proper job. Jesse Pinkman can kill people relatively easily and it may be argued that the journalist wanted to covertly convey the same opinion about Charlie Alliston. Another possible reading is that 'braking good' is a patronising phrase to explain in the simplest terms to cyclists that braking is good. These interpretations are possible subtle messages that the headline may convey to some readers even if perhaps the journalist did not mean to imply either of them. It is nevertheless significant that the headline for an article concerning a fatal collision that happened under unusual circumstances, and a trial in which the defendant was criticised by the judge for showing no remorse, refers intertextually to a well-known tv series whose protagonists lead a very unusual life and kill a vast amount of people without hesitations nor repentance.

Lastly, the headline of the tabloid article included in our analysis is 'Cyclists! On your bike if you have no licence' (Cowing 2018) uses an ironic tone to address cyclists—once again a reified group—and implies that the lack of a licence is the problem to be solved, thus ignoring the fact that 80% of adult cyclists also possess a driving licence (DfT 2012) and are hence as prepared as other drivers about road rules. As we shall see, the whole article employs a conversational style to create what Fairclough (1989, 184) describes as 'simulated equalization' whereby the text producer is perceived as on an 'equal footing with the reader through choices of expressions readers may make themselves'. The headline can also be read as an intertextual reference to the phrase 'get on your bike' attributed to Norman Tebbit, a conservative British politician who in 1981 stated that when his own father was unemployed in 1930s he got on his bike and looked for work. The phrase had been revived a short time before the publications of Cowing's article when on 21 February 2017, during a parliamentary debate, the former health Labour minister and keen cyclist Lord Hunt told Lord Tebbit to get on his bike as a reply to his complaint about the traffic in central London being slowed down 'by the barricades that have been put up to assist cyclists, who also get in the way on the main carriageways' (HL Deb 2017).

4.3 Discursive Moves in Discrimination Against Cyclists

We shall now move to the application of van Dijk's discursive moves described above to the four articles under investigation to show the links between the denial of racism in the press and the stigmatisation of cyclists. The first move identified by van Dijk is positive self-presentation. This is part of the ideological square (van Dijk 2016, 74), which consists in the emphasis of positive self-description and negative other-description while minimising our negative properties and their positive properties to create ideological polarisation. The four articles analysed presented various examples of positive self-presentation.

The opening lines in *The Times* present the typical structure 'I have nothing against ***, but...' and a few moves constructing an 'us vs them' ideology can be observed.

> This newspaper has for many years trumpeted the economic, environmental and physiological benefits of cycling, particularly in towns and cities. Even the most enthusiastic advocates of two-wheeled transport, however, cannot be blind to the persistent disrespect shown by a minority of cyclists to both the highway code and other users.

The positive presentation pre-empts criticism by foregrounding that the author and the whole newspaper know well the benefits brought by cycling in cities. Still, the choice of the quoting verb 'trumpeted' could signal a distancing move by the journalists who may not fully agree with this view. In these few lines there are three references to cyclists, but the only one in which the journalist uses the word 'cyclists' is the one that refers to *their* negative properties, while the two positive aspects employ the nominalised form 'cycling' and the convoluted phrase 'the most enthusiastic advocates of two-wheeled transport'. This is consistent with the rest of the editorial, as a positive reference to cyclists is substituted with 'enthusiasts for pedal power' and another negative one refers to 'certain cyclists' (see also Sustrans 2019; Fincham 2007).

In the second sentence, we can identify the two following discursive moves identified by van Dijk: denial and counter-attack and moral blackmail. The journalist talks about 'the persistent disrespect' presenting his opinion as an irrefutable fact and thus introducing the reversal move which implies that 'we' are the real victims and that the real enemy are 'a minority of cyclists' but also 'the most enthusiastic advocates' who 'cannot be blind', which subtly suggests that actually they *are* blind. The metaphor of blindness closely reminds us of the 'ideologically blinkered anti-racists' of which van Dijk talks and also introduces the move of moral blackmail implying that *we* (the journalist and the newspaper) are not blind, but look at things as they are and state 'the truth'.

As already mentioned, Shariatmadari in *The Guardian* tries to have a more positive attitude towards cyclists, but in the attempt to strike a balance, some of his words reinforce stigmatisation against cyclists. His opening lines show his positive self-presentation:

> I love cycling. I am not a bike nut, and those who are would laugh at the secondhand lump of steel that costs more to service than I paid for it.

The journalist positions himself among people who love cycling, but significantly avoids stating that he is a cyclist. Throughout the text the word 'cyclist' is mostly employed to discuss the negative properties of cyclists, even if—as we shall observe later—he gradually moves from referring to cyclists as 'they' to employing 'we'. In fact, the second sentence appears to suggest that the normality for 'real' cyclists is being 'bike nuts' who spend too much money on their bicycles, while he does not possess the technical skills to avoid paying more to service that the price of the bicycle itself. This second sentence has to do with the double stigma described by Aldred (2013), Shariatmadari here positions himself as on the 'non-competent' side of the spectrum, thus probably gaining the sympathy of people who cycle but would not describe themselves as 'cyclists'.

With a similar approach, Cowing, the author of the article published in the *Scottish Daily Mail*, tries to create a bond with readers by ironically describing her positive properties as a driver:

I DID not pass my driving test first time around. We need not get bogged down in the details of just how many attempts were made, but suffice it to say that by the time I passed I had spent more time with my infuriated driving instructor in the previous calendar year than members of my own family.

Looking back, though, I don't regret it. That it took me 12 months and [redacted] times to pass meant that by the time I was unleashed on the road by myself I was a relatively competent and experienced driver. One who knew how to reverse around a corner (a mysterious manoeuvre never used before or since) and leave the right amount of room for a cyclist.

This introduction, with its tongue-in-cheek style, has the twofold goal of creating a feeling of equality between the journalist and her readers and implying that all drivers, thanks to the fact they own a driving licence, are 'competent and experienced'. The informal and metaphorical expressions 'bogged down', 'unleashed', the vivid image of the 'infuriated driving instructor' and the ironic introduction of '[redacted]' to hide the number of times Cowing took a drive test all contribute to the creation of a positive image of the writer that readers can identify with.

The last example of positive self-presentation we shall observe, from *The Independent*, blends this move with that of moral blackmail.

Cycling, like Nicola Sturgeon, is a sacrosanct subject that produces furious reactions. Last year, when I dared to suggest that not all cyclists were considerate road users, that many pedestrians feel threatened by aggressive riders and that the building of special lanes for cyclists had brought traffic in London to a standstill, I received a torrent of abuse which lasted for months.

We observe here, once again, the use of the nominalised form to talk about positive aspects of cycling and the use of the noun 'cyclists' to describe their negative properties. The reason why Street-Porter chooses to compare cycling to the First Minister of Scotland Nicola Sturgeon requires readers to be aware of events that had taken place in 2015, when Street-Porter reported being threatened for having joked about Sturgeon on her Twitter account (Street-Porter 2015). The use of the adjective sacrosanct is ambiguous, as it may appear as a positive self-presentation move, but it can also be interpreted as implied criticism levelled at those who exaggerate by thinking it is sacrosanct.

The discourse then jumps straight to the moral blackmail argument, as Street-Porter denounced the 'torrent of abuse' she received, implying she is bravely doing her job and fighting censorship by publishing this article. This move opens the following one, which is the second discursive move identified by van Dijk, denial and counter-attack.

As explained above, according to van Dijk this discursive move is enacted in three steps and the third step points at 'the real racists' and 'the real enemy' (104). In the Alliston-related articles analysed here we see that the real enemy is identified with 'the cycling lobby' (Street-Porter), 'a militant minority' (*The Times*), 'the cycling campaigners' (Cowing), with the exception of Shariatmadari's article which does not play around this confusion and only refers to the problems of cyclists in general. If we observe the actions that are attributed to these 'real enemies' we see that the language employed is reminiscent of informal everyday conversations criticising cyclists in general.

> The cycling lobby is vociferous, single-minded and determined to shout down anyone who can't see that cycling brings you closer to godliness. (Street-Porter)
>
> On occasions, the refusal of a militant minority to respect traffic lights, stop signs and pavements results in tragedy. (*The Times*)
>
> All this is fine, say the cycling campaigners. Cyclists are safe! Let's not put barriers in the way by making them sit a test! Pfft, if we have a cycling test, we should have a pedestrian test, too! (Cowing)

We notice here the implied ironic criticism of Street-Porter and the reference to godliness and, in Cowing, the imitation of the campaigner's reaction through the insertion of the onomatopoeic 'pfft!' These discursive choices testify how these articles implicitly tend to invite the reader to share their views against cyclists in general by imitating the informal conversations these readers may have. Similarly, *The Times* lists the three 'mistakes' commonly attributed to cyclists. The interesting aspect here is that the reader is likely to infer that Kim Briggs' death was due to one or more of these behaviours, while in fact Alliston was condemned for riding an illegal bike without a front brake.

The third move identified by van Djik is moral blackmail, as explained above, in three of the articles analysed—*The Times*, *The Independent* and *The Scottish Daily Mail*—the reference to moral blackmail was blended with the positive self-presentation of the journalist or the newspaper. Each journalist, more or less explicitly, stated that their choice of writing something against cyclists would be met with resistance and hostility, while, in fact, their articles are all published in mainstream newspapers and many of the readers' comments, when present, agree with the journalists' opinions.

Subtle denials, i.e. 'expressing doubt, distance or non-acceptance of statements or accusation by others' (van Dijk 1992, 105) are also present in the articles under scrutiny. An example is Shariatmadari's statement on the perception of safety that, he argues, characterises all cyclists. Earlier in his article he states that 18mph is 'fast enough to kill yourself or someone else'. From a discursive perspective, in order to see why this statement actually consists in a subtle form of discrimination against cyclists, it is useful to contrast it imagining the same statement applied to motorists. In fact, in a different article published in *The Guardian*, the lawyer Martin Porter explained that

> The Highway Code gives a typical stopping distance of 12 metres for a car driving at 20mph, suggesting that if Briggs had stepped into the path of a "slow" moving car, the driver would not have been able to avoid her. [...] On any objective view, 18mph is a cautious speed and on a busy London street matching the speed of other traffic, rather than going much slower, is a wise precaution. Any suggestion that it is too fast applies a peculiar double standard and would potentially criminalise many riders. (Porter 2017)

Shariatmadari's article focuses only on speed on bicycles and the readers are not invited—and hence are unlikely—to compare the accident between Alliston and Briggs to an accident between a motorist driving at 18mph and a pedestrian or a cyclist. In his article, in fact, he invites cyclists to fear the consequences of their cycling activity and, consequently, he is contributing to increase fear of cyclists in those who do not cycle. As Horton (2007, 146) explains, reproducing concerns about the way cyclists behave—in a context dominated by 'socially and ecologically destructive automobility'—is a classic example of scapegoating, as the representation of cyclists as 'deviant' has the effect of confirming the use of cars as 'normal'. All this, he adds, is facilitated by the low numbers of people cycling, which means a small number of people will take and defend the point of view of cyclists.

It is the scapegoating attitude, present in all the articles under scrutiny, that makes each journalist introduce various forms of mitigation and concessions. These are typically followed by further attacks against cyclists, that is the defense and offence move identified by van Dijk (1992, 106–107). For example Shariatmadari states that cars are far more dangerous than bikes, but the semantic construction of his sentences is particularly significant:

> It is a stark reminder for cyclists that bicycles, even ones as shoddy as mine, can injure and kill. Many do recognise this, and take great care when they ride. And it goes without saying that cars are far more dangerous than bikes. They are responsible for all but a tiny handful of traffic deaths.
>
> That doesn't mean cyclists don't have a specific problem, though. (Shariatmadari 2017)

First of all, the phrase 'it goes without saying that' is normally employed to introduce information that is obvious and known by the addressee. The obvious information here is that 'cars are far more dangerous than bikes', but in fact the way in which this and the other articles under scrutiny portray cyclists and the use of bicycles will induce readers to perceive the opposite, as fatalities caused by car drivers are perceived as normal and inevitable while fatalities or collisions caused by someone riding a bicycle are framed as a new phenomenon that is growing, causing important problems and that should be addressed by creating new legislation. As a consequence, the phrase 'it goes without saying that' appears to diminish the importance of the statement it introduces, it is the fact that cycling advocates keep repeating, hence it is quoted, but it is not the focus of Shariatmadari's article. The phrase allows mentioning it in passing, conceding that cycling advocates are also right, and swiftly moving to the next point.

That phrase is not the only item that deserves attention. In the following sentence the pronoun 'they' is a reference item with more than one possible referent, i.e. it could refer to either cars or bikes (Halliday and Hasan [1976] 2013, 310–314). In the same sentence the journalist uses the metaphor 'a tiny handful', which belongs to the realm of cooking and thus refers to food. Discussing metaphors in religious discourse, Charteris-Black (2014, 200) states 'in these cases there is a reversal of the normal expectation that the consumption of food and drink provides satisfaction as instead it proves to be the cause of damage and even destruction. Therefore food and drink metaphors are often used for the pragmatic effect of heightening the contrast between a negative and a positive evaluation.' As a 'tiny handful' conveys an image of scarcity and the reader will associate the low number of deaths to the use of bicycles, at first the ambiguous pronoun 'they' could be interpreted as referring to the item 'bikes' rather than 'cars', on closer reading, the recipient would then notice the particles 'all but'.

The point of this analysis is to show that, when Shariatmadari touches upon the contrast between the number of deaths caused by people driving cars and people riding bikes, his syntax becomes blurred and convoluted, inserting in the same sentence an ambiguous pronoun, an unusual metaphor and an adversative conjunction. This complex sentence consists in the concession he makes to the cause of cycling advocates, before the next discursive move, which is again an attack on cyclists 'That doesn't mean cyclists don't have a specific problem, though'. The conjunctive 'though' creates an adversative relation, meaning 'contrary to expectation' (Halliday and Hasan [1976] 2013, 250), and this sentence is also characterised by a double negation, implying that someone thinks that cyclists do not have a problem and that someone can be identified with cyclists themselves or 'the cycling lobby'. The use of the negatives reduces the force of what is being stated, namely that, 'even if cars are responsible for the vast majority of traffic deaths, cyclists have a specific problem'.

Even if apparently sensible and sensitive, the whole argument presented in the article under scrutiny is problematic, as its point is to state that drivers are tested, controlled and they are aware that cars are dangerous, while cyclists are not aware of the fact they can also kill. In reality, the number of traffic deaths caused by people driving cars tells the opposite (Baker 2019), and the idea ('illusion' according to Shariatmadari) that 'we [cyclists] can't really be that much of a menace to ourselves or to others' is coherent with what traffic deaths statistics prove. This explains why the passage quoted above is convoluted, as its function is to concede the counterargument to the article while drawing attention away from it. As the aim of Shariatmadari is to write in favour of cycling and in favour of a safer and more pleasant street environment, what we have observed testifies to the presence of hegemonic discourse favouring car mobility, which is so powerful that it also affects articles that want to promote velomobility.

Mitigation, the next discursive move identified by van Dijk, can take the form of euphemisms, downtoning or other circumlocutions and usually generates a concession which, when observed within the context of the whole text, proves to be only apparent. Street-Porter's article provides an example:

> Only two people a year are killed by cyclists, against around 400 fatalities caused by vehicles, so this is not a growing problem requiring a special Act of Parliament. [...] I agree the current laws are antiquated, and do not reflect the increasing danger to all road users and pedestrians – 100 cyclists were killed in 2015, up 12 per cent on the previous year, and 3,000 were injured. There are other aspects of cycling which demand attention – it should be a legal requirement to wear a helmet, as drivers and passengers have to wear seat belts. Children under a certain age should not be allowed to cycle on city roads without an adult accompanying them. Bells or horns should be mandatory. Anyone who rents a Boris Bike should have to hire a helmet, and be prepared to submit to a breath or drugs test.

In the first two sentences reported here, the journalists appear to concede the arguments that advocacy groups would raise in response to the issue of people killed by people who cycle and quotes statistics that make the article appear balanced and based on facts. The lexical choices are revealing: those who kill (i.e. do something negative) are 'cyclists' and 'vehicles'—removing the agency and responsibility of people who drive those vehicles, but those who wear seat belts (i.e. do something positive) are 'drivers' and 'passengers'. The apparent concessions based on statistics are not there to actually argue in favour of better protection of people who cycle, but to argue in favour of legislation that will make 'them' more like 'us' drivers, listing a number of platitudes and focusing in particular on the introduction of helmet mandatory laws, which have been demonstrated to be detrimental (see Chapter 3).

Mitigation is also present in *The Times*, in a construction which reminds of the typical racist statement 'the majority of black people are fine, but some of them...'

> The majority of enthusiasts for pedal power are law abiding. Some, sadly, are irresponsible, reckless and downright dangerous.

This commonplace, which could obviously be applied to citizens in general, when applied to a specific group (be it cyclists, blacks, homosexuals, women, Muslims…) results in an attack towards that community and discursively implies that the negative characteristics attributed pertain exclusively to that group in society. This brings us to the final discursive move identified by van Djik, defence and offence, which in his work focuses on the attacks against the anti-racists. This move is often linked to the reversal according to which 'we are the real victims' and van Dijk observes the amalgamation of comparisons, metaphors and ironic reference (1992, 107–108). In this passage from the *Scottish Daily Mail* we can observe various components of this discursive move.

> And meanwhile, it's drivers who get it in the neck. One genius recently suggested the introduction of a new part of the driving test 'where part of the theory sees a driver viewing the road from a cyclist's perspective'. Really? How is that going to stop cyclists knocking down pedestrians? There is no doubt that there are great benefits to cycling: it's green, it's efficient, it's healthy, and those little baskets on the front are simply adorable. But if our roads are to be swamped with cyclists, it seems only common sense that they learn how to use them safely and responsibly.

Drivers, framed as 'the real victims', 'get it in the neck', a colloquial expression based on a metaphor that means 'to be severely reprimanded or punished' (OED 2019). Once again we notice, like in *The Times*, the contrast between 'cyclist' associated with something negative and 'cycling' to something positive (albeit in this case with an ironic undertone). Irony is also present in the reference to 'one genius', creating a false debate about the usefulness of the proposed new driving test in order to prevent collisions between cyclists and pedestrians while the change to the test was obviously envisaged to protect people who cycle from collisions with motorised vehicles. The provocative unanswered questions are followed by an apparent concession, listing 'great benefits', which introduces the usual negative conjunction 'but' and another metaphor 'are to be swamped'.

Like in Street-Porter's article, we observe the use of 'our roads' in which 'our' is likely to be understood as meaning that roads do not belong to cyclists. The passive verb form 'are to be swamped' implies that some unspecified agents—probably the institutions who promote increase in cycling—will 'swamp' our roads with cyclists, who thus become objects without agency. Charteris-Black (2006, 570) groups the verb ‘to swamp’ with other flood metaphors which are typical of anti-immigration discourse (wave, flow, flood) and more in general testify the existence of the conceptual metaphor IMMIGRATION IS A NATURAL DISASTER (572). For our purposes, his further observation on this metaphor is even more significant, as Charteris-Black (570) explains that the reason why metaphors referring to liquids are preferred is to be found at a still less conscious level: liquids are related to the primary conceptual metaphor CHANGES ARE MOVEMENTS as, by their nature, they move around. He explains that lack of control over change is lack of control over movement and suggests that ‘“liquid” or “water” metaphors are not so much about controlling the physical movement of people, but controlling the rate of historical change in a society over time.’

4.4 Concluding Remarks

What are we to make of the observations reported in this chapter? Mainstream media reports concerning a trial against one person who killed another person in an unusual, unlikely and unexpected way reveal discursive strategies that have much in common with the strategies employed by other journalists to save their face against (expected) accusations of racism. The process observed in both situations is that one characteristic of the individual(s) being criticised for their actions is foregrounded, and this shifts the focus from the individual(s) to an imagined group or community of people who share that specific foregrounded characteristic and thus the group, not the individual, becomes the object of criticism. This generalisation against a community can be judged negatively, so the journalists need to prevent negative judgment by denying that they are making an overgeneralisation and by pre-emptying criticism. The further step is that the people who denounce and try to solve the problems experienced by those who belong to that imagined group or community are portrayed in the media as ‘the real enemy’.

What is the focus, why is this parallel discrimination significant? The parallel discursive features observed can contribute to the understanding of the excess of hate suffered and reported by people who advocate in favour of better cycling infrastructures. Just like racism, bikelash and deliberate aggressive behaviour against pro-cycling parades are driven by irrational and misplaced fears of change. In order to understand this, we need to enlarge the focus on a wider perspective. Shaer's quotation (2011) reported in Wild et al. (2018) significantly employs another water metaphor 'bike lanes have gone from simple strips of pavement festooned with green and white paint to sponges for a sea of latent cultural and economic anxieties', which also in this case is connected with Charteris-Black's above-mentioned observation about social change. Wild et al. provide a useful reflection on the shades of discontent that bikelash embeds and offer practical solutions to address them and improve the effectiveness of bike schemes. But a deeper, further reflection is required if we want to recognise the true nature of the opposition against people who cycle, and especially against those who want to be able to cycle to more places and more often.

Poli (2011, 164–166) envisages three approaches to traffic policy, 'traditional', 'moderate reform' and 'revolution'. In a traditional approach the planning relies on revealed preferences, thus reinforcing the current trends, following the preferences revealed by the market. In terms of traffic policy it means that 'if roads are filled with cars, it means that we have to build more roads to respond to the demand' (164). The second option of moderate reform tries to invert the growing trend in car use by increasing public transportation. It tries to respond to the existing transportation needs by simply changing part of the mobility choices. This option is welcomed by many environmentalists and, Poli explains, matches the sustainable development idea, as 'in the attempt to involve traditional industry in environmental projects, the sustainable growth environmentalists have become lost in convenient negotiations and have forgotten the real goals' (166). The revolutionary approach questions the very need of mobility as we know it and includes strategies of traffic reduction. Poli (168) compares the need to reduce traffic by reducing the need for car use to the similar problem of waste management and the need to reduce the excess of waste production by addressing the origin of the problem, i.e. excessive packaging.

A mobility shift from driving a private car to cycling as a main means of transport implies questioning and changing everyday habits and revisiting one's mobility needs, often reducing distances and reducing the amount of goods one needs to transport. The most obvious example is that of people who think they need a car because they regularly need to reach the large-distribution cheaper shopping centre located in a decentralised area designed to be reached by car exclusively—and bring heavy goods back home. Changing one's shopping habits may result in giving up private car ownership, thus saving much more money than that saved by choosing the 'cheaper' shopping centre. But this change challenges one's identity as a consumer, and, given the vast amount of effort the industry of advertising invests in constructing our personal identities as profoundly linked to what we buy and where we buy it (including of course our identity as drivers driving a specific car model), resistance to any change to this kind of market-induced self-perception is understandable and natural.

The linking of anti-cyclists discourse and racist discourse allows to bring to the fore the same attitudes of resistance to social change. This perspective can help cycling advocates—and policymakers who genuinely want to increase cycle use—to understand the origin of those negative attitudes and thus respond accordingly rather than getting lost in rational explanations regarding practicalities. On a deeper level, the implication is that if the promotion of cycling is embedded within what Poli labelled a 'moderate reform' approach, its effectiveness will be very limited, as the origin of the problem will not be tackled. In order to be effective and generate a highly significant modal shift, the promotion of cycling needs to be 'revolutionary', in Poli's terms, and question the real origin of the problem. It needs to be part of a holistic approach that aims to make cycle use more possible and more likely, not only by creating safe infrastructures but by questioning and revising the market-induced mobility needs at work.

References

Aldred, R. (2013). Incompetent or Too Competent? Negotiating Everyday Cycling Identities in a Motor Dominated Society. *Mobilities, 8*(2), 252–271.

Aldred, R. (2016). Stakeholders, Policy and Media. In M. Bliemer, C. Mulley, & C. Moutou (Eds.), *Handbook on Transport and Urban Planning in the Developed World Institute of Transport and Logistics Studies*. Sydney, NSW, Australia: University of Sydney.

Baker, P. C. (2019, October 3). Collision Course: Why Are Cars Killing More and More Pedestrians? *The Guardian*. https://www.theguardian.com/technology/2019/oct/03/collision-course-pedestrian-deaths-rising-driverless-cars. Accessed 23 November 2019.

Charteris-Black, J. (2006). Britain as a Container: Immigration Metaphors in the 2005 Election Campaign. *Discourse and Society, 17*(5): 563–581.

Charteris-Black, J. (2014). *Analysing Political Speeches: Rhetoric, Discourse and Metaphor*. London: Palgrave Macmillan.

Cowing, E. (2018, April 21). Cyclists! On Your Bike If You Have No Licence. *The Scottish Daily Mail*.

DfT. (2012). THINK! Cyclist. https://www.gov.uk/government/publications/think-cyclist. Accessed 10 June 2019.

DfT. (2017). *Reported Road Casualties Great Britain 2016*. https://assets.publishing.service.gov.uk/government/uploads/system/uploads/attachment_data/file/668504/reported-road-casualties-great-britain-2016-complete-report.pdf. Accessed 20 May 2019.

DfT. (2019). *THINK! Campaigns*. https://www.think.gov.uk/campaigns/. Accessed 12 June 2019.

Fairclough, N. (1989). *Language and Power*. London: Longman.

Fincham, B. (2007). Bicycle Messengers: Image, Identity and Community. In D. Horton, P. Rosen, & P. Cox (Eds.), *Cycling and Society* (pp. 179–195). London: Routledge.

Furness, Z. (2010). *One Less Car: Bicycling and the Politics of Automobility*. Philadelphia: Temple University Press.

Gilligan, V. (2008). *Breaking Bad*. Santa Monica, CA: High Bridge Productions.

Goddard, T., Ralph, K., Thigpen, C. G., & Iacobucci, E. (2019). Does News Coverage of Traffic Crashes Affect Perceived Blame and Preferred Solutions? *Evidence from an Experiment Transportation Research Interdisciplinary Perspectives*. https://doi.org/10.1016/j.trip.2019.100073.

Grierson, J. (2017, August 23). Ex-courier Convicted for Mowing Down Woman on His Track Bike. *The Guardian*. Retrieved online https://www.theguardian.com/uk-news/2017/aug/23/ex-courier-charlie-alliston-convicted-for-mowing-down-kim-briggs-on-his-track-bike.

Halliday, M. A. K. (1994). *An Introduction to Functional Grammar*. London: Edward Arnold.

Halliday, M. A. K., & Hasan, R. [1976] (2013). *Cohesion in English*. London and New York: Routledge.

Hendrickson, R. (2000). *The Facts on File Dictionary of American Regionalisms*. New York: Facts on File.

HL House of Lords. (2017). *February 21 Debate (vol 779 col 201)*. Retrieved from https://hansard.parliament.uk/Lords/2017-02-21/debates/9A234351-46B5-4D97-B679-A798F1C8476E/AirPollution. Accessed 12 June 2019.

Horton, D. (2007). Fear of Cycling in Horton. In D. Horton, P. Rosen, & P. Cox (Eds.), *Cycling and Society* (pp. 133–152). London: Routledge.

Johnson, M. (2014, September 17). Why Do We Still Hear People Joke About Hitting Cyclists? *The Conversation*. https://theconversation.com/why-do-we-still-hear-people-joke-about-hitting-cyclists-31214. Accessed 16 May 2019.

Kilgarriff, A., Baisa, V., Bušta, J., Jakubíček, M., Kovář, V., Michelfeit, J., et al. (2014). The Sketch Engine: Ten Years On. *Lexicography, 1*, 7–36. http://www.sketchengine.eu.

OED Online. (2019). Oxford University Press. www.oed.com.

Poli, C. (2011). *Mobility and Environment: Humanists Versus Engineers in Urban Policy and Professional Education*. Heidelberg: Springer.

Porter, M. (2017, August 23). Motorist Would Not Have Landed Cyclist's 'Wanton and Furious Driving' Charge. *The Guardian*. Retrieved online https://www.theguardian.com/environment/bike-blog/2017/aug/23/motorist-would-not-have-landed-cyclists-wanton-and-furious-driving-charge. Accessed 20 June 2019.

Ralph, K., Iacobucci, E., Thigpen, C. G., & Goddard, T. (2019). Editorial Patterns in Bicyclist and Pedestrian Crash Reporting. *Transportation Research Record, 2673*(2), 663–671. https://doi.org/10.1177/0361198119825637.

Rissel, C., Bonfiglioli, C., Emilsen, A., & Smith, B. J. (2010). Representations of Cycling in Metropolitan Newspapers-Changes Over Time and Differences Between Sydney and Melbourne, Australia. *BMC Public Health, 10*(1), 371.

Rothman, L. (2013). Breaking Bad: What Does That Phrase Actually Mean? *Time*. http://entertainment.time.com/2013/09/23/breaking-bad-what-does-that-phrase-actually-mean/. Accessed 10 June 2019.

Shariatmadari, D. (2017, August 24). After Kim Briggs's Death, Cyclists Must Realise That They Are Traffic Too. *The Guardian*. Retrieved online https://www.theguardian.com/commentisfree/2017/aug/24/kim-briggs-cyclists-traffic-bike-charlie-alliston.

Spagnolo, C. (2016, January 22). Lecce, piomba con l'auto sui ciclisti e fugge: uno è morto, un altro è grave. *Forse una lite, un fermato Repubblica*. Retrieved online https://bari.repubblica.it/cronaca/2016/01/22/news/auto_pirata-131829166/.

Sparks, M. (2011, March 1). Driver Ploughs into Critical Mass Ride in Brazil. *The Guardian*. Retrieved online https://www.theguardian.com/environment/bike-blog/2011/mar/01/diver-brazilian-critical-mass-ride.

Street-Porter, J. (2015, April 17). Janet Street-Porter: Death Threats and How My Joke About This Ice Cream Exposed the Sick Heart of the SNP. *Mail Online*. Retrieved online https://www.dailymail.co.uk/news/article-3042846/JANET-STREET-PORTER-Death-threats-joke-ice-cream-exposed-sick-heart-SNP.html.

Street-Porter, J. (2017, September 22). Cyclists Are Rude, Abusive and Uncivil—But They Do Deserve More Protection on Our Roads. *The Independent*. Retrieved online https://www.independent.co.uk/voices/cyclist-london-cycling-charlie-alliston-kim-briggs-helmets-vehicles-pedestrians-a7961551.html.

Sustrans. (2019, May). *Active Travel in the Media: Exploring Representations of Walking and Cycling in UK and Scottish Online News.* Retrieved online https://www.sustrans.org.uk/sites/default/files/file_content_type/susr1682_active_travel_in_the_media_srp_v2.0.pdf.

The Guardian. (2019). Do Cyclists Think They're Above the Law, and Does It Even Matter? https://www.youtube.com/watch?v=DBP2LTQxqZ8. Accessed 20 June 2019.

The Times. (2017, August 25). Braking Good. Retrieved online https://www.thetimes.co.uk/article/braking-good-cyclist-charlie-alliston-convicted-of-wanton-and-furious-driving-after-death-of-kim-briggs-5k2mg0hdb.

Urry, J. (2004). The 'System' of Automobility in Theory. *Culture & Society, 21*(4/5), 25–39.

van Dijk, T. A. (1992). Discourse and the Denial of Racism. *Discourse and Society, 3*(1), 87–118.

van Dijk, T. A. (2016). Critical Discourse Studies: A Sociocognitive Approach. In R. Wodak & M. Meyer (Eds.), *Methods of Critical Discourse Studies* (3rd ed., pp. 62–85). London: Sage.

WHO. (2018). *Road Traffic Injuries.* https://www.who.int/en/news-room/fact-sheets/detail/road-traffic-injuries. Accessed 12 June 2019.

Wild, K., Woodward, A., Field, A., & Macmillan, A. (2018). Beyond 'bikelash': Engaging with Community Opposition to Cycle Lanes. *Mobilities, 13*(4), 505–519.

CHAPTER 5

London Mayor's Transport Strategy

Abstract In this chapter, the discursive strategies employed in the 2018 Mayor's Transport Strategy and the Cycling Action Plan for London are observed and analysed. Linguistic notions are outlined and their relevant features for critical discourse analysis explained—transitivity, nominalisation, modality, lexical choices and agency. Corpus Linguistics (CL) and Corpus-Aided Discourse Analysis (CADS) are briefly introduced. In the texts under scrutiny, the absence of a narrative of conflict and the inclusive ways of speaking about road users show a clear attempt to stir away from an 'us vs- them' approach. A section is dedicated to nudging strategies and their ideological implications observing how the issue of free choice is dealt with in the corpus analysed. The chapter also reveals the pervasive paradigm of the market society, which is prominent in both documents, and shows how systems thinking can help us understand and reframe the relation between the geobiosphere and the economic system.

Keywords Nudge · Transitivity · Nominalisation · Agency · Growth

As explained in Chapter 2, this book investigates the discourse of cycling following two main threads, that of marketisation and that of an 'us vs. them' narrative. The previous chapter has focused almost exclusively on the latter; this chapter observes how both threads emerge in institutional documents published by the Mayor of London and Transport for

M. C. Caimotto, *Discourses of Cycling, Road Users and Sustainability*, Postdisciplinary Studies in Discourse,
https://doi.org/10.1007/978-3-030-44026-8_5

London in 2018. As we shall see, the analysis of discursive strategies reveals that the authors of the documents have taken great care to avoid a narrative of conflict and have employed inclusive ways of speaking about road users. On the other hand, we can also observe the presence of the pervasive paradigm of the market society, which is employed as an argument in favour of changing mobility patterns because this can help achieve 'Good Growth'. This investigation focuses in particular on the 'Mayor's Transport Strategy' (Mayor of London 2018) published in March 2018 and the Cycling Action Plan (TfL 2018a) published in December 2018.

5.1 Linguistic Features and Corpus-Aided Methods

Before moving to the examination of the texts, for readers who are less familiar with discourse analysis we shall introduce the main linguistic features under scrutiny and provide a brief explanation of what corpus-aided discourse analysis methods consist in. When we analyse a text, it is useful to start from identifying all the verb forms. We can then observe transitivity, that is, broadly, who does what to whom and how. This approach is derived from Halliday's seminal work *An Introduction to Functional Grammar* (1994); differently from traditional grammatical approaches—which distinguish verbs that take objects ('Mary broke the glass') from those that do not ('John slept')—Halliday emphasises that the grammar of a language is a system of options from which speakers and writers choose according to social circumstances and transitivity plays a key role in 'meaning making' in the language. Through these choices, it is possible to background or omit responsibility, as agency and the responsibility for actions can be left implicit. Transitivity analysis is thus a powerful tool for analysing not only what is *in* texts but also what is *absent* from them (see also Machin and Mayr 2012, 104–136; Simpson et al. 2019, 73–80 for an introduction to the analysis of transitivity).

While we identify verb forms, we can also notice nominalisations—verb processes that have been transformed into noun constructions—a transformation whereby the people doing the action are removed, hiding both the agent and the affected. Nominalisations can remove any sense of time and, as they become a thing, they can be counted, described and classified—while causality becomes a second concern. As nominalisations become a thing, they can become new participants and, through repeated use, become stable entities that enter common usage. A typical example is 'globalisation' which conceals the fact that globalisation is not a stable thing in itself but rather the result of a process derived from specific political and economic decisions. Nominalisations can create a sense of action, as the process is still there, but they avoid agents, times and specificity through simplification. Overall, texts become more dense and compressed and this helps make solutions appear simpler and feasible (see Machin and Mayr 2012, 137–162 for an introduction to the analysis of nominalisations).

The next aspect that emerges as relevant in the texts we are going to analyse is that of modality. According to Fairclough, 'modality is a major dimension of discourse, and more central and pervasive than it has traditionally been taken to be' (1992, 160). His approach (1992, 158–162) is also derived from Halliday, expanding from the traditional grammar view that associates modality to modal auxiliary verbs ('must', 'may', 'can', 'should' and so forth) and broadening it to include any unit of language expressing the writer's personal opinion of what they say, such as hedging ('I believe/think/suppose'), modal adverbs ('probably', 'possibly', 'obviously') and their equivalent adjectives (e.g. 'it is likely/ clear that…'). Modality, Fairclough explains, can be 'subjective'—the author tells us who is expressing a degree of affinity with what is being said ('I believe', 'they argue', 'scientists have proved')—or 'objective', in which case it is not clear whose perspective is being represented. In some cases the statement does not appear to express an opinion but an actual fact, presented as a categorical assertion not marked for modality. Fairclough points out that 'the use of objective modality often implies some form of power' (1992, 159).

Of course, lexical choices represent an important part of text analysis: they are the most salient part on which readers who do not have a background in linguistics are likely to concentrate. When we pay attention to which words have been selected from the possible available options, we take into account the connotations of the words employed. Establishing connotations objectively, especially if we are analysing texts we agree with, can prove a difficult task, as what we choose to focus on, or the connotations we attribute to words, will be influenced by our views. Hence starting our analysis from the observation of transitivity, nominalisations and modality can help us distance ourselves from the text. Apart from explicitly defining the researcher's own sociopolitical position as recommended by van Dijk (2001, 96), we can also make our analysis more objective by using corpus-aided techniques.

Corpus linguistics (CL) consists in 'dealing with some set of machine-readable texts which is deemed an appropriate basis on which to study a specific set of research questions'. As CL 'is not a monolithic, consensually agreed set of methods and procedures for the exploration of language', the balance can move either towards a mainly quantitative approach or towards a qualitative one and these should be seen as the extremes of a continuum rather than as binary opposites (McEnery and Hardie 2012, 1). A CL approach aims to provide a method allowing us to study language within the context of the scientific method, granting that the research results are accountable, falsifiable and replicable (Leech 1992). But, of course, it is necessary to bear in mind that a corpus is a selection of language, which is in principle non-finite, hence accountability can only be claimed as far as our selected corpus is concerned (McEnery and Hardie 2012, 15). The way in which CL is employed in this chapter consists in what is generally referred to as a 'corpus-assisted' or 'corpus-aided' approach (CADS), which Partington, Duguid and Taylor (2013, 10) define as a 'set of studies into the form and/or function of language as communicative discourse which incorporate the use of computerized corpora in their analyses'. One of the main differences between CL and CADS is that with the latter approach analysts usually read their texts, while CL is mostly employed to work with amounts of texts that would be too big to read in full (see Marchi and Taylor 2018, 3–6 for a discussion on the complementarity of qualitative and quantitative).

In particular, the Mayor's strategy document is investigated here through the use of WMatrix (Rayson 2008), an online-based tool that allows analysts to compare the text(s) under scrutiny against what is known as a 'reference corpus', which is larger and used to reveal what is statistically prominent or absent in the corpus being investigated. This process in CL is mainly employed to reveal keywords, but WMatrix also allows the semantic tagging of a corpus, thus identifying which key themes are more prominent and which ones are missing. The software that performs the tagging is called USAS (short for UCREL[1] Semantic Annotation System). As Koller et al. (2008, 144) explain, 'USAS

[1]UCREL is in turn the acronym for the University Centre for Computer Corpus Research on Language, which is based at Lancaster University (UK).

assigns semantic domain tags, which are pre-defined in the underlying lexicon, to the types in a corpus; and Wmatrix compares the frequency of the tags thus allocated to a large external reference corpus — subsets of the British National Corpus (100 million words) — to ascertain tag keyness via statistical significance. Significance is measured using log likelihood, with a threshold value of 6.63 for $p<0.01$'. This method is useful to obtain a quick overall view of what is debated in the text(s) and from which perspective(s); moreover it can also point towards the right direction when we are looking for the metaphors that are significantly frequent, although the software was not designed with this purpose in mind. This can be achieved by paying attention to semantic domains that are unexpected in the texts that we are observing. All the information retrieved from the corpus will still require to be evaluated by the researcher, so corpus-aided methods help achieve more objectivity, but the researcher still plays a prominent role at the beginning of the research, when the research questions are posed and the corpus is collected, and at the end, when the quantitative results are interpreted.

5.2 Three Key Themes

On the web page where the full document is provided, the Mayor's Transport Strategy (Mayor of London 2018) is introduced as follows:

Three key themes are at the heart of the strategy.

1. Healthy Streets and healthy people

Creating streets and street networks that encourage walking, cycling and public transport use will reduce car dependency and the health problems it creates.

2. A good public transport experience

Public transport is the most efficient way for people to travel over distances that are too long to walk or cycle, and a shift from private car to public transport could dramatically reduce the number of vehicles on London's streets.

3. New homes and jobs

More people than ever want to live and work in London. Planning the city around walking, cycling and public transport use will unlock growth in new areas and ensure that London grows in a way that benefits everyone.

From a linguistic perspective, agency is particularly relevant here. The only active verb phrase in which the agent is human (people) is 'want to live and work'. Other actions are performed by agents that are either personified through metaphors or are actions themselves in nominalised forms: streets and street networks *encourage*, creating streets *will reduce* car dependency, car dependency *creates* health problems, a shift *could reduce*, planning *will unlock* growth, London *grows* in a way that *benefits* everyone. Also 'the strategy' is personified through the use of a metaphor, as it has *a heart*. Transitivity-related choices background the actors who will actually create the streets or plan the city. They also background the fact that *people* are car dependent and *people*, often those same people, suffer from health problems. In line with this, we notice that it is the *number of vehicles* that could be reduced, not the number of people driving cars. Of course, in reality these two possible phrases correspond to the same result, it is only a matter of choosing, from the options available, how to express the idea that the aim of the strategy is to make people use their cars less. This discursive strategy, which as we will see is prominent throughout the whole document, allows the author to reduce the possible resistance from car-users and is an effective way of avoiding a common pitfall often found in materials that promote a shift from car to bicycle.

Promotional materials to invite people to cycle more often frame drivers negatively and cyclists positively, implying—or explicitly stating—that drivers are responsible for the environmental damage and are thus morally inferior to cyclist. It is the case of the 2019 'funny insults' campaign series of four posters released by Bike for Brussels (2019) in which 'cyclists' are represented by cute animal or monster-like creatures in the foreground while 'motorists' are small cars in the distant background from which huge speech balloons emerge with insults like '*bouge-toi, fils de bobo anti CO2*' (which roughly means 'move it, son of a radical chic anti CO2') and the tagline '*traitons les cyclistes comme ils méritent*' ('let's treat cyclists as they deserve'). Another version of this narrative portrays cyclists as smarter, as they have understood the advantages of cycling in traffic-congested cities while drivers stubbornly stick to their cars, missing their chance to save money, lose fat, get there earlier. For instance, the 2006 Manchester campaign recited 'burn calories, save cash, get there on time' against a picture of a road with a bike lane marked 'fast lane' and a car lane marked 'fat lane' (Manchester FOE 2006).

People who identify as drivers understandably react negatively to these narratives, listing all the reasons why they need to use a car, criticising cyclists for their attitude of being 'closer to godliness' (see Street-Porter 2017) and arguing that they are not stupid or stubborn, but the circumstances do not allow them to cycle, and that cyclists simply don't understand these needs because they belong to an elite/don't have a job/are too young/are retired/don't have a family, etc., or, worse, cyclists don't care about others and only want everyone to use a bicycle because they want to impose their choices. The campaign released by Transport for London (TfL 2019), 'Walk. Cycle. Discover.' avoids comparing car drivers and cyclists, but rather addresses a person who can choose between driving or walking and cycling. The video version lists all the possible positive things that the experience of cycling or walking will provide, while the poster promoting cycling simply promises 'A new route. A new London' with an offer image in which the viewer is invited to imagine the thoughts of a woman holding a bicycle and heading towards a sun-lighted, inviting cycle path along the river. Her everyday clothes and the Holland-style bicycle with a straw basket depict slow cycling. While the London campaign avoids the pitfalls of the Manchester and Brussels ones described above, more could be done in terms of conveying new worldviews embracing the protection of ecosystems, of all forms of life and the fostering of social justice.

All the documents related to the Mayor's Transport Strategy avoid employing road users categories (pedestrians, cyclists, drivers) thus avoiding the pitfall of the 'us vs. them' narrative that we have observed in the previous chapter. Through lexical choices and discourse strategies, the texts reframe the people who move around London, all are *Londoners* or *people* who can engage in different actions. Active mobility is referred to by using the verb-derived nouns 'walking and cycling', which reinforce the sense of action that contributes to the positivity the document attributes to active mobility. Public transport is referred to through the nominalised phrase 'public transport use' and the action of 'driving' is carefully avoided and conveyed through the nominalised phrase 'car dependency'. We can see how in this case, the removal of agency and responsibility aims to remove from the minds of the readers the categories of cyclists, drivers and pedestrians and group them all under the inclusive notion of *people* or *Londoners*. The fact that 'car dependency' conveys passivity has a double function: it creates a contrast with the sense of activity conveyed by the verbs—reinforcing the idea that not being active is not healthy—and removes the negative moral judgement against people who drive, as here they do not choose to drive, they are not morally responsible for their choices, it is rather the administration who should provide alternatives in order to help them reduce or overcome their dependency.

If we observe modality in the introduction quoted above, we notice that car dependency *will* be reduced, growth *will* be unlocked while the number of vehicles on London's streets *could* be reduced. This lower modality reveals the difficulties arising from the fact that even the best infrastructure and public transport will not grant an automatic reduction in car numbers, as, in order to achieve that, people need to modify their behaviour, a result that is particularly difficult to obtain and requires time. The sentence 'More people than ever want to live and work in London' is an example of what Fairclough (1992, 159) describes as 'objective modality', as it is a categorical statement not attributed to a source and one which has actually been challenged after the Brexit referendum (Tapper 2019). The underlying message that is conveyed with the third point—i.e. growth is good—frames more people, new houses and new jobs as indisputable good news (see also Halliday 2001 for the notion of 'growthism'). From a lexical point of view, this positivity is conveyed through the repeated use of 'new', 'grow/growth', 'unlock' and 'benefit'.

5.3 Positive Discourse Analysis

As mentioned in Chapter 2, one of the aims of this book—in line with Stibbe's view (2018, 165)—is to reveal discourse strategies that can help create 'new discourses to base society on'. We shall thus start by focusing on how the Mayor's Strategy contributes to create or reinforce a worldview that will improve people's lives through changes to mobility. Let us observe the text under scrutiny through WMatrix, listing the top 20 overused items, that is, the themes (semantic domains) tagged through USAS and sorted on the log-likelihood value, which is a statistic used for significance testing. For the corpus-assisted analysis, the 'accessible version' of the document was employed, which consists in a text version in large print that also provides information concerning the visual part of the document, adding to the original versions captions like 'Front cover image Photo of Hoxton Overground Station entrance area'. While this introduces a slight change compared to the standard publication, these additions can be considered a valuable way of redressing the fact that a corpus-assisted investigation can only focus on text and leaves images out, even if, of course, their presence plays an important part in communicating the messages. In Table 5.1, under column O1 we can find the observed frequency in the Mayor's Strategy while in O2 we find

Table 5.1 Prominent themes in the Mayor's Strategy compared to the BNC Sampler written

USAS tag	*O1*	*Relative freq.*	*O2*	*Relative freq.*		*Log-likelihood*
Vehicles and transport on land	3326	5.20	2171	0.22	+	11405.04
Helping	1211	1.89	4225	0.44	+	1510.91
Wanted	1043	1.63	5233	0.54	+	825.88
Evaluation: good	737	1.15	2905	0.30	+	802.95
Important	658	1.03	2803	0.29	+	652.24
Time: future	907	1.42	4846	0.50	+	651.67
Places	955	1.49	5888	0.61	+	534.50
Geographical names	1759	2.75	14,502	1.50	+	495.66
Moving, coming and going	1287	2.01	10,157	1.05	+	410.67
Using	435	0.68	1965	0.20	+	399.53
Inclusion	366	0.57	1624	0.17	+	344.23
Giving	679	1.06	4474	0.46	+	332.98
People	502	0.79	2896	0.30	+	317.21
Measurement: area	103	0.16	93	0.01	+	313.70
Green issues	131	0.20	225	0.02	+	289.15
Safe	117	0.18	166	0.02	+	288.33
Money: cost and price	290	0.45	1254	0.13	+	282.04
Change	578	0.90	3939	0.41	+	263.84
Healthy	112	0.18	195	0.02	+	245.15
Quantities: many/much	526	0.82	3576	0.37	+	241.36

the observed frequency in the Written British National Corpus (BNC) Sampler, which is employed as reference corpus. Table 5.1 shows the top 20 prominent themes in the Strategy. The Mayor's Strategy is a small corpus of 63.932 words (known as 'tokens' in CL) and the reference corpus is 968.267 tokens.

When using these results we must bear in mind that the corpus we are analysing actually consists in a single (albeit quite long) text. For sake of completeness, the numbers are provided, but, as explained above, the results from the corpus-analysis software are employed mainly as a source of inspiration for a deeper qualitative analysis. The first line tells us that the topic of vehicles and transport is statistically much more prominent than in the reference corpus for written English, which is hardly surprising given that the text is about transport. We shall focus later on the specific words employed to discuss the topic. The second item is less predictable and when we look at the words listed under this semantic tag, we find many verbs like 'support', 'enable', 'help', 'benefit', 'encourage' and 'promote'. These indicate that the approach the document aims to convey is one of collaboration between the institutions and the citizens, avoiding a discourse of imposition. The themes of Places and Geographical names are prominent because of the many occurrences of 'boroughs', 'city' and 'local' for the former and 'London', 'Londoners' and the various parts of London frequently mentioned. The next item worth observing is 'Using' as it confirms the statistical prominence of 'use' referred to car use and public transport use that we have already discussed when we observed the introduction. The 'Inclusion' domain is prominent because of the frequent use of 'including', 'include' and 'inclusive', which reinforces the discourse of collaboration.

The next tag 'Giving' is actually misleading, as its prominence is due to the high presence of the word 'emission(s)'. Following that, we find 'People', due mainly to 314 occurrences of the word 'people'. This confirms the strategic choice of referring to those interested in and affected by the proposed strategy as one group rather than different categories. It is clearly a discursive strategy employed to prevent and reduce potential conflict and to convey the idea that everyone is interested in the same fundamental goal of living in a healthier and safer environment. Moreover, it conveys that everyone has to be included and helped to achieve what is necessary in order to obtain that goal. This reduction of conflict is confirmed by one of the most significant underused items (thus absent from the table): tokens tagged as 'Warfare, defence and the army, weapons' only occur 7 times in the Mayor's Strategy, with a relative frequency of

0.01, while they occur 3152 times in the reference corpus, with a relative frequency of 0.33, LL 342.46. Moreover, none of the 7 occurrences is a war metaphor, confirming that nowhere in the text we find anything related to the narrative of the 'war on the roads' (see Walker 2019). Concerning the remaining themes, it is worth mentioning that the prominence of 'Measurement: Area' is mainly due to the 97 occurrences of 'space(s)', while the prominence of 'Money: Cost and price' may be considered misleading as it is related to the list of interventions followed by the specification of high-cost/low-cost. The prominence of 'Change' is significantly due to the high presence of the word development, occurring 118 times with 0.18 relative frequency in the singular form, which in some cases is employed as a synonym of growth in general and in other cases refers to new developments as areas for people to live in.

Table 5.2 shows a selection of the keyword list generated by WMatrix (always using the BNC Written Sampler as a reference corpus, and a LL threshold of 6.63). The items were manually selected to show their position in the full keyword list, and thus their prominence in the text. The items selected confirm the discursive strategies that had already been observed through qualitative analysis. The last three items are shown to highlight their scarce presence in the text.

Table 5.2 Selected list of keywords

Order in full list	*Item*	*O1*	*Relative freq.*	*O2*	*Relative freq.*		*Log-likelihood*
2	Public_ transport	336	0.53	24	0.00	+	1695.97
3	Transport	392	0.61	117	0.01	+	1646.94
8	Cycling	225	0.35	8	0.00	+	1183.09
13	Walking	221	0.35	50	0.01	+	976.71
15	Vehicles	175	0.27	18	0.00	+	856.20
16	Londoners	154	0.24	4	0.00	+	819.94
26	Freight	123	0.19	25	0.00	+	553.04
29	Car	185	0.29	190	0.02	+	533.70
35	Cycle	106	0.17	22	0.00	+	475.05
47	People	314	0.49	1086	0.11	+	395.36
52	Traffic	94	0.15	51	0.01	+	341.40
59	Vehicle	66	0.10	12	0.00	+	301.73
76	Cars	77	0.12	59	0.01	+	249.77
381	Drivers	13	0.02	8	0.00	+	45.44
382	Cycles	10	0.02	2	0.00	+	45.07
632	Cyclists	5	0.01	1	0.00	+	22.54

Table 5.2 shows the predominance of active travel, through the use of 'public transport' and the reference to walking and cycling, which are referred to using—*ing* participle forms to convey a feeling of action. People—not categorised according to road user categories, but rather grouped as 'people' or 'Londoners'—are at the centre and are sometimes contrasted with cars, as in

> London's streets should be for active travel and social interaction, but too often they are places for cars, not people. (12)
>
> It [This simple aim of a shift away from the car] will reconnect communities by creating places where people are prioritised over cars. (19)
>
> Streets that are busy with people, rather than cars, are safer. (41)

The lines reported at the bottom of the table show the very few occurrences of road user categories and it is worth adding that 4 out the 5 occurrences of 'cyclists' actually come from the image descriptions that are only present in the accessible version we used for machine reading. Thus particular care has been taken to avoid the use of the word 'cyclist' and this is confirmed by observing the draft document which had been released for public consultation in June 2017. All the (few) occurrences of 'cyclists', but one, that were present in the draft (D) were substituted in the final (F) document, here are two examples:

> (D) While car use is lower than in outer London, the area presents the greatest immediate opportunity to reduce it even further by transforming facilities for cyclists. (28)
> (F) [...] transforming facilities for people cycling. (30)
> (D) The interactions of connected and autonomous vehicles with pedestrians, cyclists and conventional vehicles. (262)
> (F) The interactions of connected and autonomous vehicles with people walking, people cycling and conventional vehicles. (284)

5.4 The Issue of Choice

From a discursive point of view, the Mayor's Transport Strategy (Mayor of London 2018) does not leave any space for the possibility that anyone could still want to choose using the car over other more active mobility choices, even when provided with the best possible infrastructure for walking and cycling. In fact, if we observe the way in which the notion of choice is constructed, we notice that there is an underlying assumption that choice can and should be controlled by an external agent. This is actually in contrast with the very notion of choice:

> But we also need to look at how we can fundamentally change the way people choose to move around our city. (6)

> However, congestion on public transport and the street network is unlikely to reduce without a focus on behaviour change that enables and encourages people to choose to walk and cycle for short and medium-length journeys. (253)

> By improving the environment for walking, cycling and public transport users, these will become the default modes of choice for travel rather than the use of private vehicles. (265)

As the choice of active travel is framed in the text as the right choice for one's health, Mulderrig's analysis (2018) of 'nudge' tactics from a critical perspective can help us investigate the discursive strategies deployed in the document. Her analysis focuses on the *Change for Life* campaign, a social marketing campaign targeting children and their parents, using commercial marketing techniques to influence lifestyle behaviours—eating, drinking and exercising. She highlights how the increasing popularity of 'nudge' tactics—an expression coined by Thaler and Sunstein (2008)—'helps legitimate and instantiate neoliberal political rationalities by privatising (both structurally and morally) responsibility for public health care' (Mulderrig 2018, 39). However, she underlines that even if associated with the Conservative government, forms of soft paternalism were already in place under the preceding Labour government (44). She explains the bipartisan appeal of nudge on a number of levels: financial, political, cultural and ideological.

Firstly, nudge is cheaper than regulatory and fiscal alternatives, thus making it an attractive complement to austerity policies. Secondly, its efficacy is largely unprovable, while at the same time allowing governments to show they are taking action. Thirdly, nudge draws on technologies of discourse proven to succeed in the marketplace and thereby resonates with a generation of people increasingly oriented to such consumer-based forms of relation and identification. Finally, the ideological principles underpinning nudge claim to address the fundamental challenges of neoliberal politics (45).

Does all this mean that it is wrong to encourage people to eat more healthily, to exercise and prefer active mobility choices? It does not, but the above argument is complex and clarification might help: the idea that people can and should be encouraged to favour walking and cycling over motorised private means of transport is certainly in line with this book's ecosophy (see Chapter 1). One of the ideas underpinning this research project is that if we are healthy we feel better and we are thus more likely to lead a happier life. Hence being provided with more and better possibilities to live and move in less polluted, less traffic-congested and safer environments is desirable and those in charge should act in order to achieve this result. Still, if we want policies to work on a large scale, to be accepted and implemented effectively, it is necessary to play devil's advocate and carry out a fine-grained critical analysis revealing any underpinning—and potentially detrimental—discourse.

In fact, this problem was raised in the consultation report provided on TfL website (TfL 2018b):

> While participants liked the idea of a less car-dependant London they were wary about the role of individual behaviour change and believed a major shift was needed. They questioned whether the strategy had fully accounted for entrenched habits and social norms that underlie car dependency, and whether the scale of change required is achievable. (41)
>
> Key requests included: Need to better recognise the challenge of creating behaviour change from the car to active travel choices. (49)
>
> Key suggestions included: Behavioural change • A strategy to encourage mode shift to active modes for Londoners who are dependent on motor cars […] • Proposals should be more radical and promote a society where fewer people actually own cars. (75)

TfL recommendations specified that no changes to the strategy were required in response to these comments, inserting the links to the documents in which their 'extensive research and analysis on mode shift and propensity for people to switch modes' are reported (76). While evaluating or discussing their methods and findings goes beyond the scope of this book, some discourse features are worth mentioning. One of the main topics investigated concerns the appeal of using a car and the evaluation of which people are more likely to accept the mode shift. Under the section 'Healthy Streets and healthy people' of the consultation report we find a paragraph worth quoting in full (emphasis in the original):

Three quarters of car trips could be made by a more sustainable mode, but many people are reluctant to change.

There remains great potential to deliver mode shift from the car to more sustainable modes across London, and particularly in outer London – 74 per cent of all Londoners' car trips and 93 per cent of car trips under 2 km could feasibly be made by an alternative mode. Some people are more amenable to change than others – the challenge is that the greatest amount of car use happens in places where people are most committed to travelling by car and least willing to change. Delivering mode shift requires changing people's preferences. In practice, realising the potential that has been identified would mean car travel becoming less appealing and other modes more appealing in terms of their cost, convenience and freedom from stress. (TfL 2017, 7)

Once again we can observe here how transitivity has been employed to distance people from their actions: the passage opens with an existential 'there' and the existential verb 'to be' is employed four times; and the actions of the authorities to generate the required change are expressed through impersonal forms—'to deliver', 'delivering', 'changing' and 'realising'. The two final sentences could be rephrased more explicitly as 'In order to make people change their mode of transport we (the Mayor of London/ TfL/ Authorities in charge etc.) need to change what people prefer. In practice, if we want to realise the potential that we have identified [i.e. get the number of people potentially willing to shift transport mode to actually shift], we need to make car travel less appealing and other modes more appealing'. The use of the definitional verb 'mean', a verb usually employed to equate two mutually substitutable linguistic items like 'café means coffee' (see Fowler and Kress 1979), paired with the use of the verb 'becoming' contribute to the strategy of agency deletion.

Agency is backgrounded to create distance from what citizens will need to do—or will be 'nudged' into doing—and from what those in charge will implement in order to make people change their preferences. This discursive strategy is most likely deployed because the authors expect readers to react negatively to the ideas presented. The document from which this passage is taken is publicly available on the TfL website but it was not envisaged and promoted to be read as widely as the Mayor's Strategy. In fact, when we compare the two documents we notice that the latter avoids mentioning the idea that car travel will be made less appealing and only focuses on the positive side of the matter. The following concordance shows all the 23 occurrences of the word 'appealing' and the words that precede and follow each of them in the text.

Left	*Node*	*Right*
nd indirect, meaning there is no	appealing	alternative to car use for many.
public transport become the most	appealing	and practical choices for many mor
transport options accessible and	appealing	to all Londoners is the key to red
walking and cycling are the most	appealing	choices for getting about locally
t option for longer journeys more	appealing	. Improved bus routes particularly
means that individual streets are	appealing	places to walk, cycle and spend t
s, and a city that is made up of	appealing	streets and places will attract bi
perates as a whole. London needs	appealing	walking environments in every neig
in comfort and safety. It needs	appealing	cycling environments and a strateg
ransport connections that provide	appealing	alternatives to car use. The way
m a Tube station every day. More	appealing	walking environments will encourag
nts will also make cycling a more	appealing	option. Cycling allows journeys t
g Making walking and cycling more	appealing	to all Londoners requires a big ch
the delivery of Healthy Routes.	appealing	environments are required to encou
Londoners and also become a more	appealing	option than car use by: a) Impro
oth easier to understand and more	appealing	to existing and future customers.
nomous vehicles make car use more	appealing	and easier to do, people may walk
d sustainable travel options more	appealing	than shared mobility services. Ma
on streets are already relatively	appealing	places to walk, cycle and spend t
significantly improved to provide	appealing	and convenient alternatives to car
ependent on providing new or more	appealing	alternatives. Compared to people
. The walking environment will be	appealing	so that 3 to 5 million more trips
alternatives are not available or	appealing	. Car-lite development: A housing

This approach appears to be in contrast with a point clearly stated in the Supporting Evidence document 'It is unlikely that interventions targeted only at improving the alternatives would be sufficient to persuade this group to change mode' (TfL 2017, 90).

As Baslington (2008, 111) explains, when authorities deploy 'travel awareness' campaigns and individual marketing approaches, they are dealing with individuals who are car dependent and may potentially change their behaviour, but they are not addressing the effects of a social dependency on cars. She states that an alternative is to tackle the problem by 'reducing the availability and desirability of cars'. Lyons (2004) warns policymakers against the idea that individuals are 'rational, optimizing agents' (502) and highlights how people are, on the one hand, creatures of habit who tend to be resistant to change but, on the other hand, they are also incredibly versatile and adaptable to change, even if change is always perceived as painful, whether it is self-imposed or imposed by others, which is perceived as even more painful (503). In his analysis he also focuses on the role of the media in highlighting and exploiting their readers' resistance to change, reporting the negative coverage of the introduction of congestion charging to enter Central London when it was introduced in February 2003 and how the media showed lack of interest when the scheme proved successful.

When we observe the overall discursive approach followed in drafting and then publishing the Mayor's Strategy, we can recognise a clear attempt to focus on the positive aspects and the positive outcomes. The positive effects for the collective interest and the consequent advantages for individuals are foregrounded while the natural human resistance to change is backgrounded and omitted. Change is presented as chosen by individuals, not imposed, and thus discursively constructed as more acceptable and even actively searched. This, together with the choice analysed above of avoiding narratives of conflict and the categorisation of road users, amount to discursive strategies that aim to move away from the main trends observed in most cycling-promoting documents. By observing the statements in the supporting documents and the changes between the draft and the final version, we can reasonably argue that these discursive choices are conscious. The third discursive aspect that we shall observe stirs away from these two and adheres to mainstream discourse: everything related to commerce and growth is presented as fundamental, growth is framed positively and other possible views are totally silenced. By observing the other documents available, we have no reason to think that foregrounding business and growth as positive simply amount to a discursive strategy, on the contrary they appear to be the core of the vision.

5.5 The Issue of Business and Growth

In the texts under scrutiny, the positivity of growth is unquestioned. Halliday (2001, 194) explains that the grammar of our languages contain growthism, as resources are shown to be unlimited (oil, water, air, energy, etc., are uncountable nouns) and humans occupy a special position (see s/he vs. it and the reluctance to attribute agency to non-humans). Moreover, as Halliday points out, the 'growth word' in terms of contrast like 'big' and 'small' is always the neutral term, as we ask 'how fast is the car' (not how slow) and 'how high is the building' (not how low). What we observe in our analysis of the texts is even more explicit. An example is present in the contrast created between ordinary people's journeys and business-related journeys. As already mentioned, people driving cars are referred to through the nominalised form 'car dependency', in a similar way business-related journeys are referred to as 'freight' or 'freight trips' (see Table 5.2). While people's trips made by car are envisaged to be replaced by trips made by public transport or active travel—business-related trips are to be organised more efficiently, but the document never mentions the idea of *reducing* the need to move or to transport goods. In fact, freight and commercial journeys are often associated with the adjective 'essential', which is an example of objective modality: which journeys are to be considered essential is not specified—an objection also raised in the consultation report, where a definition of 'essential journey' is requested (TfL 2018b, 87)—hence the decision is left implicit and could either mean that it is up to people to evaluate the level of importance of their journeys or it could depend on policies that will be introduced in the future.

The unquestioned positivity of growth is intertextually linked to other publications by the Mayor of London and, like in the Healthy Streets case, the reference is signalled by the capital letters of the expression 'Good Growth', as in the following passage:

> Growth is good for London, and it is important that all of the city's current and future residents feel its benefits. As the city grows, it must also become a better place to live in – London's growth must be 'Good Growth'. (26)

We can observe here the nominalisation of 'growth', which transforms a process into a fact. While a process could be altered either by political choices or under other external influences, the use of the nominalisations removes agency and removes the idea that alternatives could be chosen and pursued. The process of growing is framed as something happening naturally and independently from the people living and administering the city. We can also observe the use of the existential verb 'to be' (growth is good)—presenting an opinion as a fact—and the presence of high modality ('is', 'is important', 'must become' and 'must be'). One of the main documents in which the Mayor outlined his Good Growth notion is *A City for All Londoners* published in October 2016 (Greater London Authority 2016). As Spinney underlines, mobilities scholarships 'have failed to critically engage with the relationship between cycling and the neoliberal state' (2016, 451). Both his work and that of Popan analyse London as an example of the 'conceptualisation of cycling through a narrow productivist framing, both materially and discursively' (2019, 99).

The centrality, relevance and unquestioned positivity attributed to growth in the Mayor's Strategy are in line with the way in which, apart from a very small number of exceptions, 'sustainability' and 'sustainable development' are framed and envisaged by politicians, policymakers and laypeople. The Mayor's Strategy strives to counterbalance the widely held view that the introduction of environmentally friendly policies will undermine the economy and foregrounds the idea that a healthy economy can only flourish in a healthy environment. Albeit, the healthy economy projected in the Strategy is based on growth, and the idea that growth is positive is discursively constructed as an undisputable fact. This unbridled faith in growth is at the very centre of what keeps humanity from finding effective solutions that put an end to the ongoing destruction of ecosystems (see also Popan 2019, 181–185).

Comparing the Mayor's Transport Strategy (Mayor of London 2018) with the Cycling Action Plan (TfL 2018a) published at the end of the same year, we can observe consistency in the discursive strategies deployed: conflict narratives are avoided, change is framed as chosen by people and the importance of business, money and economic growth are foregrounded. The first of eight chapters is about 'Cycling's role in London's success' and the centrality of money-related aspects is prominent. While from a quantitative point of view this is not particularly significant, the choice of opening the document by highlighting the economic benefits is a strategy to adhere to the mainstream marketised discourse (Mautner 2010). What follows is a selection of examples (emphasis added):

> [...] the benefits of cycling, which include better physical and mental health, *improved productivity* and the freedom to travel *cheaply* around the city. (9)

> Enabling more Londoners to cycle is not just about helping people get around the Capital quickly and easily. It also assists in *creating the conditions for London's economy to thrive* by keeping the city moving. Cycling helps our streets move more people at the busiest times, *making the city function more efficiently for businesses.* (14)

In Chapter 2, drawing on Lakoff's work (2010) we outlined the tragic hypocognition that makes us humans conceive ourselves as detached from nature and the environment of which we are part. We need to change the ways in which we talk about humanity and the environment and in order to do so, we need to change the way in which we *think* about them, not an easy task given that we use language to think and our languages make us conceive the environment as detached. A possible way out to help us approach problems, without being drawn back by the mental models that fail us, is Systems Dynamics (Meadows 2008; Sterman 2012). Unfortunately, Sterman explains, most efforts made by individuals and institutions to enhance sustainability are directed at the symptoms and not at the causes and systems (any system) will respond to any change introduced with what is known as 'policy resistance', that is the existing system will tend to react to change in ways that we had not intended when we first designed the intervention (a few examples are road-building programs designed to reduce congestion that ends up increasing traffic or antibiotics that stimulate the evolution of drug-resistant pathogens—for a longer list and further explanation see Sterman 2012, 24). Systems Dynamics allows us to calculate scientifically the way in which a complex system will react to change and to account beforehand for what we usually describe as 'side-effects'. Side effects, Sterman argues, 'are not a feature of reality but a sign that the boundaries of our mental models are too narrow, our time horizons too short' (24). As Gonella et al. (2019) explain:

> As long as we consider the geobiosphere as a sub-system (a resources provider) of the human-made economic system, any attempt to fix environmental and social problems by keeping the business as usual, i.e., the mantra of economic growth, will fail. The reality tells us the reverse: geobiosphere is not a sub-system of the economy, economy is a sub-system of geobiosphere. As systems thinkers know, trying to keep alive at any cost the operation of a sub-system will give rise to a re-arrangement of the super-system – the geobiosphere – that will self-reorganize to absorb and make ineffective our attempt, then continuing its own way. (Gonella et al. 2019)

This brings us back to Mautner's point about society being embedded within the economy rather than the opposite (2010, 16). Of course, in a city where people are healthier, feel happier and live better the local economy will also benefit from the improved living conditions. But framing growth as a certain and certainly positive fact while framing cycling as a means to aid growth is problematic. It could reasonably be argued that there are political needs to make the Strategy welcomed by all-powerful groups and that compromise is necessary in order to achieve results. The point of the analysis presented here is to underline how this utilitarian way of thinking may well contribute to achieving results in the short term, but also contributes to the hypocognition that prevents us from being able to *think* differently and envisage effective, long-term solutions.

5.6 Concluding Remarks

What are we to make of the discursive strategies deployed in the Mayor's Transport Strategy? It is a document that can certainly be considered innovative and ambitious. It represents an important milestone: it actively tries to change the discourse of cycling by addressing some of the communicative pitfalls in which both advocacy groups and institutions tend to fall. From a linguistic perspective, the discursive strategies employed to convey a positive feeling, making readers feel that the results will be achieved smoothly, focusing on the outcomes and highlighting all the positive effects, backgrounding or omitting the efforts that the introduction of changes in people's everyday lives and habits will require. It also reassures readers about the most widely dreaded effect of environmental policies, i.e. negative effects on the economy, and foregrounds that growth is positive, that its centrality is not being questioned and it will actually thrive thanks to the changes introduced.

The first two linguistic aspects that we have investigated are clearly discursive strategies to frame what is being discussed in ways that will influence how readers react to the text, namely reducing the conflict that often arises when discussing relationships between different road users. Moreover, the resistance that people are likely to show towards imposed changes concerning their mobility habits is reduced by making them believe that they will choose active mobility once its self-evident advantages will be visible and the choice will have been made feasible. The third discursive aspect—confirming and reinforcing the mantra of growth—is the most problematic. On the one hand, it must be acknowledged that the changes outlined are already quite disruptive of the status quo, hence if the document had dared to introduce ideas that are still considered 'extremist' by most people—namely the fact that growth has limits, that we live on a finite planet and that 'the question is not if growth will cease, but when and how' (Sterman 2012, 31)—the whole project would have been ostracised. On the other hand, it is necessary to recognise that by choosing to reaffirm the mantra of growth and to frame the Transport Strategy as functional in order to achieve 'Good Growth', the document adds itself to the vast number of attempts appreciated and supported by people who genuinely care about environmental issues but do not see, or choose to ignore, the intrinsic oxymoronic contradiction of searching for 'sustainable growth', and considering endless growth with finite resources an achievable—and even desirable—goal.

References

Baslington, H. (2008). Travel Socialization: A Social Theory of Travel Mode Behavior. *International Journal of Sustainable Transportation, 2*(2), 91–114.

Bike for Brussels. (2019). *Bike for Respect*. https://bike.brussels/en/insults. Accessed 14 November 2019.

Fairclough, N. (1992). *Discourse and Social Change*. Cambridge: Polity Press.

Fowler, R., & Kress, G. (1979). Rules and Regulations. In R. Fowler, R. Hodge, G. Kress, & T. Trew (Eds.), *Language and Control* (pp. 26–45). London: Routledge and Kegan Paul.

Gonella, F., Almeida, C. M. V. B., Fiorentino, G., Handayani, K., Spano, F., Testoni, R., et al. (2019). Is Technology Optimism Justified? A Discussion Towards a Comprehensive Narrative. *Journal of Cleaner Production, 223*, 456–465.

Greater London Authority. (2016). *A City for All Londoners*. https://www.london.gov.uk/sites/default/files/city_for_all_londoners_nov_2016.pdf. Accessed 10 November 2019.

Halliday, M. A. K. (1994). *An Introduction to Functional Grammar*. London: Edward Arnold.

Halliday, M. A. K. (2001). New Ways of Meaning: The Challenge to Applied Linguistics. In A. Fill & P. Mühlhäusler (Eds.), *The Ecolinguistics Reader: Language, Ecology and Environment* (pp. 175–202). London and New York: Continuum.

Koller, V., Hardie, A., Rayson, P., & Semino, E. (2008). Using a Semantic Annotation Tool for the Analysis of Metaphor in Discourse. *Metaphorik, 15*, 141–160.

Lakoff, G. (2010). Why It Matters How We Frame the Environment. *Environmental Communication, 4*(1), 70–81.

Leech, G. (1992). Corpora and Theories of Linguistic Performance. In J. Svartvik (Ed.), *Directions in Corpus Linguistics: Proceedings of the Nobel Symposium 82* (Vol. 65, pp. 105–122). Berlin: Mouton de Gruyter.

Lyons, G. (2004). Transport and Society. *Transport Reviews, 24*(4), 485–509.

Machin, D., & Mayr, A. (2012). *How to Do Critical Discourse Analysis*. London: Sage.

Manchester FOE. (2006). *Fast Lane Fat Lane Ad Campaign*. https://www.manchesterfoe.org.uk/loveyourbike/fast-lane-fat-lane-ad-campaign-2006/. Accessed 14 November 2019.

Marchi, A., & Taylor, C. (2018). *Corpus Approaches to Discourse, a Critical Review*. London and New York: Routledge.

Mautner, G. (2010). *Language and the Market Society: Critical Reflections on Discourse and Dominance*. London: Routledge.

Mayor of London. (2018). *Mayor's Transport Strategy*. Greater London Authority. https://www.london.gov.uk/what-we-do/transport/our-vision-transport/mayors-transport-strategy-2018. Accessed 18 July 2019.

McEnery, T., & Hardie, A. (2012). *Corpus Linguistics: Method, Theory and Practice*. Cambridge: Cambridge University Press.

Meadows, D. H. (2008). *Thinking in Systems: A Primer* (D. Wright, Ed.). White River Junction, VT: Chelsea Green Publishing.

Mulderrig, J. (2018). Multimodal Strategies of Emotional Governance: A Critical Analysis of 'Nudge' Tactics in Health Policy. *Critical Discourse Studies, 15*(1), 39–67.

Partington, A., Duguid, A., & Taylor, C. (2013). *Patterns and Meanings in Discourse: Theory and Practice in Corpus-Assisted Discourse Studies (CADS)*. Amsterdam and Philadelphia: John Benjamins.

Popan, C. (2019). *Bicycle Utopias: Imagining Fast and Slow Cycling Futures*. London and New York: Routledge.

Rayson, P. (2008). From Key Words to Key Semantic Domains. *International Journal of Corpus Linguistics, 13*(4), 519–549.

Simpson, P., Mayr, A., & Statham, S. (2019). *Language and Power a Resource Book for Students*. London and New York: Routledge.

Spinney, J. (2016). Fixing Mobility in the Neoliberal City: Cycling Policy and Practice in London as a Mode of Political—Economic and Biopolitical Governance. *Annals of the American Association of Geographers, 106*(2), 450–458.

Sterman, J. D. (2012). Sustaining Sustainability: Creating a Systems Science in a Fragmented Academy and Polarized World. In M. P. Weinstein & R. E. Turner (Eds.), *Sustainability Science: The Emerging Paradigm and the Urban Environment*. New York: Springer.

Stibbe, A. (2018). Positive Discourse Analysis: Rethinking Human Ecological Relationships. In A. F. Fill & H. Penz (Eds.), *The Routledge Handbook of Ecolinguistics* (pp. 165–178). New York and London: Routledge.

Street-Porter, J. (2017, September 22). Cyclists Are Rude, Abusive and Uncivil—But They Do Deserve More Protection on Our Roads. *The Independent*, retrieved online https://www.independent.co.uk/voices/cyclist-london-cycling-charlie-alliston-kim-briggs-helmets-vehicles-pedestrians-a7961551.html.

Tapper, J. (2019, June 29). Tired of London: Thousands Flee Capital for a Quieter Life. *The Guardian*, retrieved online https://www.theguardian.com/uk-news/2019/jun/29/tired-of-london-thousands-flee-capital-for-a-quieter-life.

TfL—Transport for London. (2017). *Mayor's Transport Strategy: Supporting Evidence*. http://content.tfl.gov.uk/mts-challenges-and-opportunities-report.pdf. Accessed 31 July 2019.

TfL—Transport for London. (2018a). *Cycling Action Plan*. http://content.tfl.gov.uk/cycling-action-plan.pdf. Accessed 18 July 2019.

TfL—Transport for London. (2018b). *Consultation Report*. https://consultations.tfl.gov.uk/policy/mayors-transport-strategy/user_uploads/mts-consultation-report-4.pdf. Accessed 31 July 2019.

TfL—Transport for London. (2019). *Active Travel*. https://tfl.gov.uk/campaign/active-travel. Accessed 21 November 2019.

Thaler, R. H., & Sunstein, C. R. (2008). *Nudge: Improving Decisions About Health, Wealth and Happiness*. New Haven and London: Yale University Press.

van Dijk, T. A. (2001). Multidisciplinary CDA: A Plea for Diversity. In R. Wodak & M. Meyer (Eds.), *Methods of Critical Discourse Analysis* (pp. 95–120). London: Sage.

Walker, P. (2019, July 8). Channel 5's 'Nonsense Will Make Me and Other Cyclists Less Safe. *The Guardian*. https://www.theguardian.com/environment/bike-blog/2019/jul/08/channel-5s-nonsense-will-make-me-and-other-cyclists-less-safe. Accessed 25 July 2019.

CHAPTER 6

Marketisation in European Documents

Abstract This chapter focuses on documents and tools created by the European Cyclists' Federation, the European Environment and Transport Ministers, the Danish public–private partnership State of Green, the World Health Organization. The texts are observed through a corpus-aided qualitative approach, investigating transitivity, nominalisations, modality and lexical choices—in particular road users labels and marketised discourse. Positive discourse strategies are highlighted. The chapter opens with a brief history of the notion of sustainable development, highlighting its ideological contradictions in order to show how these are embedded in the dominant 'pro-car neoliberal automentality' (Walks in *The Urban Political Economy and Ecology of Automobility: Driving Cities, Driving Inequality, Driving Politics*. London and New York, Routledge, 2015). The 'for what for whom' questions from Systems Thinking are blended with the discursive analysis of transitivity and agency. The corpus-aided analysis reveals the profound marketisation of the texts observed and, from these observations, starting points for reframing strategies are outlined.

Keywords Automentality · Sustainable development · Neoliberalism · CADS · European Union

M. C. Caimotto, *Discourses of Cycling, Road Users and Sustainability*, Postdisciplinary Studies in Discourse,
https://doi.org/10.1007/978-3-030-44026-8_6

In this chapter we move beyond Great Britain and investigate institutional documents addressing a European and a global audience. As mentioned in Chapter 2, according to Oosterhuis' meta-analysis (2019), the policies introduced in the last two decades in Europe, North America and Australia, have failed to generate significant increases in everyday cycling and new approaches are required. This chapter focuses on documents and tools created by the European Cyclists' Federation (ECF), the European Environment and Transport Ministers, the Danish public–private partnership State of Green (SOG), the World Health Organization (WHO). As in previous chapters, the texts are observed through a corpus-aided qualitative approach, investigating transitivity, nominalisations, modality and lexical choices—paying particular attention to the ways in which road users are referred to and to the presence of marketised discourse. In line with Stibbe's proposal (2018), positive discourse strategies are highlighted.

6.1 Limits to Growth and Sustainable Development

Before starting the actual analysis of texts, we shall sketch a brief history of the origins of the notion of 'sustainable development' and highlight how this is relevant for the analyses presented in this chapter. The first alarm came from the publication of *The Limits to Growth* report (Meadows et al. 1972) and the notion of 'sustainable development' was coined in 1987 by what came to be known as the 'Brundtland report', where it was defined as 'development that meets the needs of the present without jeopardizing the ability of future generations to meet their own needs' (WCED 1987, 37). According to Keil (2007) both documents were potentially revolutionary, as by identifying growth as a problem, they could have led to an ecological awakening revolving around anti-capitalism. Unfortunately, he explains, this was not the case, and they were actually employed to open the way to neoliberalism. Other scholars (Gonella et al. 2019, 457) underline that the Brundtland definition was already flawed as it is about 'development' rather than about a 'sustainable society' (see Brown 1981). In Keil's words:

> the Brundtland and Club of Rome proposals were subsequently recast from a critical set of potentially anticapitalist warning to a recipe for the survival of capitalism through a concrete set of measures involving ecological modernization. [...] It was precisely the internalised individual and

> community-focused responsibility of the neoliberal era that undermined both the potentially fundamental critique of capitalist development and the search for social solutions to the socioenvironmental crises that kept unfolding worldwide. (2007, 46)

From a critical discourse approach, Alexander (2009, 86) observes that 'the re-positioning of multinational corporations as promoters of "sustainable development" has proceeded so rapidly and with so much PR flak and smoke-screening that today no one appears to really know, nor do they seem to care what the concept means'. If we turn our attention to the phrase 'sustainable development', the two words are an adjective and a noun. The development consists in a nominalisation, which—as explained in the previous chapter—means that the process of *developing* is transformed into a fact, *development*, obscuring agency and the affected, removing any sense of time, creating a new nominal group that can be counted, classified and that becomes a new participant and a stable entity (Machin and Mayr 2012, 139–144). The adjective as well refers to obscured agents, as *sustainable* implies that someone will *sustain*—i.e. 'To keep in existence, maintain; spec. to cause to continue in a certain state for an extended period or without interruption; to keep or maintain at the proper level, standard, or rate; to preserve the status of' (OED Online 2019). The phrase also generates confusion around which is the status to be preserved, that of development or that of life as we know it.

In fact 'development' is another way of talking about 'growth' without necessarily entailing the idea of increase implied by the word 'growth'. Both 'growth' and 'development' can be envisaged as dead metaphors in which the economy is personified or anyway regarded as a living being. It is a way of framing them, and the ways in which they change over time, as natural processes. When looking at the definitions of 'development' (OED Online 2019), we notice that a number of them employ the spatial metaphor of 'advancement' (e.g. 'The economic and social advancement of a country, region, etc.') which is related to our, mainly western, metaphorical way of conceiving time as a movement through space (Lakoff and Johnson 2013). The metaphors of growth, development and advancement all testify to the important role these play in shaping the way we think, and of course, the same is true for 'progress'. The challenge is to reframe this deep metaphorical thinking in our brains in a way that can help us to conceive the role of humanity as less focused on moving along an imaginary line and accumulating in the process, but rather focused on staying, being and experiencing. We shall return on this challenge in Chapter 8.

Coming back to the role of neoliberalism, Walks observes the 'tight ideological symbiosis between the values promoted by automobility (individual freedom and autonomy) and the rationalities of neoliberalism' and highlights how 'the kinds of policies pursued under neoliberalism, including tax cuts, trade liberalization, the extension and guarantee of private property rights, deregulation of finance, and the attack on organized labour, have all been demanded by those capitalists benefiting from expansion of the global system of automobility' (2015, 10–11). Analysing the relationship between automobility and ideology, Walks (199–220) delineates the political contours of auto-hegemony, highlighting the growing tendency in suburban areas to support neoliberal or Christian right political parties—with the exception of poor suburbs with racial or immigrant minorities (201). He explains that, in most developed nations, post-war suburbanisation could not have occurred the way it did without the automobile and that the lifestyle he labels 'pro-car neoliberal automentality' has been internalised by people as part of one's natural rights, leading them to transform any perceived attack on the automobile as an attack on personal and political freedom.

It was during the 1970s, with the OPEC oil crisis, deindustrialisation and the publication of *The Limits to Growth* (Meadows et al. 1972) that the pro-car intellectual lobby emerged and introduced the idea of a 'war on the automobile' to refer to any policy aimed at reducing car dependency. Walks convincingly argues that the protection of the car entailed the protection and justification of suburban life, a way of life that depended on the provision of cheap energy, leading to the politicisation of energy policy (2015, 206). That decade closed with the election in 1979 of Thatcher in the UK and Joe Clark in Canada, while Reagan announced his candidacy for President of the United States—their campaigns shared the promise to allow people to maintain the way they lived and identified the problem in excessive taxes and big government. Reagan's 1980 acceptance speech is significant in this sense:

> Those who preside over the worst energy shortage in our history tell us to use less, so that we will run out of oil, gasoline and natural gas a little more slowly. Conservation is desirable, of course, for we must not waste energy. But conservation is not the sole answer to our energy needs. America must

> get to work producing more energy. The Republican program for solving economic problems is based on growth and productivity. Large amounts of oil and natural gas lay beneath our land and off our shores, untouched because the present Administration seems to believe the American people would rather see more regulation, taxes and controls than more energy. (Reagan 1980)

Investigating the promotion of velomobility in cities, Walks et al. argue that the technical and financial changes required to make cycling a significantly desired alternative mode of transport often require small investments to modify the road signage, lane markings, kerb lines and add bike racks and explicitly state 'however, it is not the economic but the political costs that have limited their implementation. Indeed, implementing even the most minor and inexpensive changes has often proven very difficult politically' (2015, 237). This chapter investigates documents that aim to address the issue on a European or even global scale. How is the discourse promoting velomobility ingrained in a neoliberal globalised system that is tied, symbolically and materially, to the automobile and to the kind of mobility needs it has created? (Urry 2004).

The linguistic analysis in this chapter draws inspiration from systems thinking literature and, in particular, Gonella et al.'s (2019) investigation concerning the narratives of technology optimism and, more in general, from the leading author of *Limits to Growth*, Meadow's *Thinking in Systems* (2008). A systems thinking approach to complex problems requires three fundamental starting points: identifying the boundaries, the end-users and the systemic purpose of the system. Hence any technological solution will require addressing two questions beforehand:

- Technology for what?
- Technology for whom? (Gonella et al. 2019, 457)

These are known as the FWFW (for what? for whom?) questions and linguists will immediately recognise how they are connected to the notions of transitivity, agents and affected. Commenting the Paris Agreement, in which growth is the ultimate goal, Gonella et al. (2019, 460) observe that 'the reality tells us that most of the new technologies are being developed to face climate change issues *provided that* at the same time the business as usual of the economic growth purpose is kept working' (emphasis in the original).

6.2 Growth and Economic Advantages in EU-Related Documents

The European Commission's webpage concerning 'Cycling Policy and Background' (EU 2019) explains that there is currently no official EU cycling strategy and lists the details of the various informal meetings and the declarations that have recognised the importance of promoting and supporting cycling. The latest of these is the 2018 'Graz declaration', which we shall analyse here together with the 2017 'EU Cycling Strategy' (EUCS) developed by the European Cyclists' Federation (ECF)—a document calling on the Commission to develop its own strategy and to provide a set of recommendations—and the 2015 Luxembourg 'Declaration on Cycling as a climate friendly Transport Mode'. Starting from the oldest one, the 1200 word declaration agreed by the EU Ministers of Transport consists in a preamble and an action plan. The preamble is divided into four sections: Innovation, Environment, Health and Financing, the first one much shorter than the others and amounting to two simple sentences: 'Cycling is a European success story. Bicycle innovation will boost jobs and growth and support EU industry through new technology and services' (EU 2015).

The central role of growth is thus foregrounded from the very beginning, and in the rest of the declaration, we find references to economic and financial aspects. In the Environment section, the second sentence describes cycling as 'the most cost effective transport mode after walking', producing 'massive positive externalities for society at little expenditure in terms of infrastructure and vehicle'. We notice the lexical choices drawn from the realm of economics, 'cost effective' 'positive externalities' 'expenditure'. Similarly, the Health section highlights the economic advantages generated by a healthy population, stating that 'employees who cycle to work claim fewer paid sick days' and that 'through congestion easing, emissions and noise reduction, public health and infrastructure cost savings, cycling benefits even those who don't practice it' (EU 2015). Foregrounding the effects on economic growth consists in a marketisation discursive strategy, as it attempts to reinforce the credibility of the declaration and make the argument acceptable by highlighting the economic aspects and by employing lexis that will be familiar to economists and perceived by other readers as grounded in economic theory. The data and facts reported are of course real, what appears problematic is that the society is embedded within the economy rather than the opposite (Mautner 2010, 16), thus improved physical and mental health, better social cohesion and life quality, safer roads, increased energy efficiency are framed as significant mainly because of the economic advantages they entail.

In a similar vein, the EUCS (ECF 2017)—titled 'Recommendations for Delivering Green Growth and an Effective Mobility System in 2030'—strives to foreground the positive economic effects that a cycling strategy coordinated at European level could bring, compared to action solely at the national, regional or local level. The report refers to the European Commission's 10 strategic priorities for 2014–2019, aiming to show how cycling 'can make a crucial contribution'. Listing these priorities, Junker's 'Agenda for Jobs, Growth, Fairness and Democratic Change' dedicates the first 6 out of 10 priorities to growth, market and economy-related issues (Junker 2014) and ECF tries to conform to the dominant discourse as in the accommodation process described by Mautner (2010). In the EUCS we find several examples where obvious societal advantages are framed as mainly significant from an economic perspective, what follows is a selection:

> Another economic benefit of cycling comes with the physical and mental health benefits of regular commuting by bike to work, namely diminished work absenteeism, hence providing significant benefits for businesses and employers. (13)
>
> A healthy lifestyle is urgently needed in view of the huge costs of a lack of physical activity to Europe's health care systems. (31)
>
> In addition to these economic benefits cycling also produces numerous positive externalities, for instance a fairer, more liveable and healthy society. (127)

In the latter we find again the economy-related phrase 'positive externalities', presenting fairness, liveability and health as nice side effects that come along with 'economic benefits'. These are framed as central, rather than the opposite. If we apply the FWFW questions (Gonella et al. 2019), in the quoted sentences cycling is promoted for the reduction of costs and it benefits those who earn money from such savings, namely employers, taxpayers and businesses.

The most recent document, the 'Graz Declaration' of 2018 (EU 2018), is about transport in general and cycling is only one of the aspects discussed. Once again, growth is framed as essential: 'This clean mobility transformation will also provide a crucial impetus for green economic growth while ensuring competitiveness for the European transport and automotive industry and creating new opportunities for businesses and green jobs' (EU 2018). Like the EUCS, the Graz Declaration refers to the HEAT tool developed by the WHO, the acronym meaning Health Economic Assessment Tool. The tool is a powerful software freely available online that allows inexperienced users to calculate economic assessments of the health impacts of walking or cycling. Based on conservative estimates, the tool calculates the economic advantages of an increase in the amount of people cycling or walking, and it can be employed to incorporate the effects on the population's health within the economic appraisals regularly carried out as part of transport planning. According to the researchers who developed the tool 'the use of economic arguments to advocate investments into policies that have a comparatively greater potential to result in health benefits is a promising strategy to win the support of other sectors', they specify that decision-makers should not use solely economic arguments, but they state that monetary terms enhance their understanding of the results (Kahlmeier et al. 2010, 120–121).

All the arguments pointing out the economic advantages of increasing cycling levels are based on rational reasoning, the studies provided by the ECF and by the WHO are envisaged as tools that can help those who want to advocate for more cycling, allowing them to support their proposals with data and evidence. Their tools and findings are the result of painstaking, complex research and are often carried out by individuals who sincerely care about increasing active mobility and who are likely to receive the kind of discursive observations exposed above arguing 'if we don't talk about money, they won't even listen to us'. And if I may add a personal note, I was utterly fascinated when in 2014—as I was getting increasingly involved in my local cycling advocacy association—the other more experienced organisers paired the yearly parade with a conference on 'Bikenomics', providing the local administrators with scientific evidence from international experts which—I naively believed—they would not want to ignore anymore.

The idea of convincing people by using the 'language they speak' and employing the arguments 'they will listen to' appears reasonable. But from a discursive perspective, as Mautner and Lakoff illustrate well, it reinforces the dominant ideology. Thus, if we assume that neoliberalism is in contrast with solutions that will reverse—or at least mitigate—climate change and the destruction of life as we know it, and if we assume that the neoliberal mentality is deeply intertwined with automobility—given that we are aware of the need to abandon the current mobility system heavily relying on private cars—then reinforcing the neoliberal mentality by arguing in favour of increases in everyday cycling, using the 'language of money', is a self-harming strategy, even when carried out with best intentions.

6.3 Lexical Choices and Metaphors from the World of Business

In order to retrieve positive discursive strategies to discuss cycling, and to identify the ones that require reframing, two different documents about cycling promotion are compared here, through a CADS approach. We shall observe the EUCS (ECF 2017) and compare it to the white paper 'Sustainable Urban Transportation. Creating Green Liveable Cities' published by the non-for-profit, Danish public–private partnership *State of Green* (SOG 2016). While both documents illustrate why an increase in cycling is desirable and how to achieve it, the EUCS is about six times longer than the white paper and reports detailed, evidence-based data and policy proposals that are compatible with the existing EU regulations. The white paper is more general, much more attractive from the aesthetic point of view, with vibrant, colourful pictures or graphs on each page. Both promote cycling, but their ultimate goal is different in that the EUCS aims to be accepted and transformed into an official EU cycling strategy with as little modification as possible, while the white paper aims to promote the successes of Denmark and to invite its recipients to collaborate with the '600 organisations, companies, utilities, research institutions, financial institutions and public sector stakeholders who can help you solve your green challenge' (SOG 2019).

The comparison thus allows us to observe the contrast between a document, the EUCS, where marketised discourse is prominent because of the attempt to conform to the dominant discourse (especially under Junker's leadership and its focus on economic growth), and a white paper, which looks like a promotional brochure and in which marketised discourse is present in a different way, as its discursive strategies are drawn from marketing. One of the basic features that make this evident is the lexical choice and frequency of the words 'problem(s)'—only

4 occurrences and all referring to theoretical, abstract problems or problems in the distant past—'challenge(s)'—12 occurrences employed to refer to problems in the present and in the future—and 'solution(s)'—44 occurrences mainly to talk about the problems that Danish experts and decision-makers have solved (SOG 2016). The EUCS highlights cycling-related problems in the EU because its aim is to call upon the EU Commission to act and implement the solutions provided in the report, while the white paper promotes (we could say 'sells') its know-how to anyone who is looking for solutions.

The EUCS document (ECF 2017) consists of 59,378 tokens and the white paper (SOG 2016) of 9235 tokens. When retrieving keywords by comparing the two texts using AntConc (Anthony 2019) with standard settings (LL 4 term, $p < 0.05$ + Bonferroni, Dice coefficient and 'all values' for the effect size threshold) the keyword lists reveal—as expected—a prominence of EU-related keywords in the EUCS and of keywords related to Denmark and to the city in the white paper. More significant for our purposes are the prominence in the EUCS of 'cycling' (+100.2), 'safety' (+43.21), 'policy' (+32.73), 'measures' (+24.59), 'should' (+22.59) and 'market' (+20.11). On the other hand, if we extract keywords from the white paper using the EUCS as a reference corpus, we observe the prominence of 'solutions' (+108.72), 'green' (+98.75), 'traffic' (+88.29), 'different' (+84.27), 'socio' (+59.8), 'more' (+50.03), 'station' (+48.92) and 'liveable' (+47.6).

The prominent keywords in the EUCS testify to its legislative nature, apart from the market-related arguments analysed above, 'safety', 'policy', 'measures' and 'should' show how the document aims to provide ready-made solutions that the EU could then transform into actual policies. As for the modal verb 'should', when it is employed in legislative documents it carries high modality, its meaning corresponding in most cases to a polite version of 'must'. The prominence of 'cycling' (1025 occurrences) is a positive strategy that brings the focus on the activity rather than the people cycling, thus avoiding the issue of categories in a conflict that we discussed in previous chapters.[1] Differently from the

[1]With the 2018 annual general meeting, ECF announced the intention of becoming the European Cycling Federation as part of its 2030 vision, thus changing the meaning of the C in the acronym from 'cyclists' to 'cycling'. This change has not taken place yet, and while on the one hand, in English, it would have positive effects and would reduce the tendency to perceive 'cyclists' as 'sport cyclists', the change could create problems in other

Mayor's Strategy (Mayor of London 2018) analysed in the previous chapter, the EUCS has 133 occurrences of 'cyclist(s)' (excluding the occurrences that are part of an acronym as in the European Cyclists' Federation), on top of these we find 9 occurrences of 'people cycling'. While the latter carries a positive prosody, we observe negative prosody even in the EUCS when it comes to 'cyclist(s)'. The most frequent nouns modified by 'cyclist(s)' are 'deaths' (6), 'protection' (3), 'fatalities' (2), while most constructions are passive, e.g. 'cyclists are put off', 'cyclists are considered', 'road schemes affect cyclists', 'systems detect cyclists', 'creating safe conditions for cyclists', 'cyclists were killed', 'include cyclists' needs' and out of 133 occurrences, in only 6 cases 'cyclists' are agents: 'are aware of danger', 'are largely unequipped', 'are willing to volunteer', 'mix with motorised vehicles' (2) and 'interact with devices'.

When we look at the keywords extracted from the white paper, we notice a positive, human-centred discursive approach. This is particularly true for the use of 'solutions', 'different', 'socio'—which is employed as a prefix for 'economic'—'more' and 'liveable'. In the EUCS we find 44 occurrences of 'economic', of which only one is 'socio-economic', while in the white paper the 19 occurrences of 'economic' are 'socio-economic' in 17 cases. The kind of argument found in the white paper is about the economy anyway, but the way it is framed shifts the focus away from it, foregrounding the societal advantages. For example, it states 'a cost-benefit analysis of a 1 km cycle journey at a speed of 16 km/h in Copenhagen rush hour shows a socioeconomic [sic] benefit of 0.22 EUR per km. In comparison, the socio-economic loss of a 1 km car journey at a speed of 50 km/h corresponds to 0.70 EUR. The total saving in cost therefore corresponds to EUR 1 per travelled km. (Source: Copenhagen Bicycle Account 2014)' (SOG 2016, 16). Looking at the list of parameters employed, it may remind us of the HEAT tool (Kahlmeier et al. 2010), in which various parameters were calculated in

languages—ECF being a federation, its name is often translated in the local languages of its members. In Italian, for example, the translation of 'cycling' with 'ciclismo' would actually reinforce the reference to sports, and the more general connotation of cycling could only be rendered with a phrase like 'andare in bici' (riding a bicycle). To avoid both 'ciclismo' and 'ciclisti', the Italian federation of local association (a member of ECF itself) was named Federazione Italiana Amici della Bicicletta (Italian Federation of Friends of the Bicycle) and later transformed into Federazione Italiana Ambiente e Bicicletta (Italian Federation of the Environment and the Bicycle).

monetary terms in order to make them comparable. But from a communicative perspective, the chart backgrounds the economic aspects and rather foregrounds the social effects, as it shows data concerning 'time, vehicle operation, health, accidents, air pollution and climate, noise, congestion, other effects' and effectively adds the prefix 'socio-' to the word 'economic' (SOG 2016, 16).

Another way of observing the discursive features of texts offered by CL is that of N-Grams, which consist in sequences of contiguous n words (types, lemmas or parts of speech). In the EUCS, AntConc reveals, among others, 57 occurrences of 'road safety' and 52 of 'added value'. The former testifies to the relevance attributed to the issue of safety, which is prominent in the document as new policies and measures are requested and the 'safety in numbers' effect is explicitly mentioned. The issue of safety is of utmost importance when planning cycling infrastructure, as the perception of danger is one of the main reasons deterring people from switching to cycling. But advocates need to be acutely aware of the fact that, discursively, talking about safety and danger will increase the recipients' perception of danger associated with cycling. Thus while it is fundamental that decision-makers make sure that safety is granted when planning cycling infrastructure, its centrality should be discursively toned down in favour of other more pleasant characteristics or expressed in different ways.

As for the repeated use of 'added value', in most cases it is part of the repeated headline 'EU added value (and/or cost of non-EU)', hence the phrase is employed metaphorically as the paragraphs actually illustrate the advantages and disadvantages of policies and measures coordinated at EU level rather than more locally. The phrase—chiefly found in the form 'value added'—is classified as 'business colloquial' and defines it as 'The benefits imparted to an undertaking by the work of a particular person or group; (more generally) additional gain or advantage'; this extended use is derived from its stringently economic meaning 'the amount by which the value of an article is increased at each stage of its production, exclusive of the cost of materials and bought-in parts and services' (the value taxed through VAT, value added tax) (OED Online 2019). The EUCS includes a number of other words and phrases drawn from economics and employed as everyday language. It is important to stress here that is not a peculiar characteristic of this specific text, but rather part of the widespread marketisation outlined by Mautner (2010).

An example of this is the use of the word 'stakeholder', which occurs 37 times in the EUCS and 11 times in the white paper. The term stakeholder became part of the mainstream academic and managerial discourse with the publication of Freeman's book *Strategic Management, a Stakeholder Approach* (1984). Stakeholder is defined as 'A person, company, etc., with a concern or (esp. financial) interest in ensuring the success of an organization, business, system, etc.,' but it is increasingly employed to talk about people's interests in general (OED Online 2019). From the perspective of economics it has been criticised for its excessive vagueness (Mitchell et al. 1997, 853–854). Two aspects of the use of the word 'stakeholder' are relevant to our analysis; firstly it is an example of marketisation as a term borrowed from the jargon of business management. Fairclough (2000) provides an analysis of the popularisation of the term by New Labour, as the phrase 'stakeholder economy' was employed during the first Blair's campaign to mean 'an economy regarded or conceived of as giving all members of society a stake in its success' (OED Online 2019). Secondly, the confusion between actors and those acted upon, as highlighted by Mitchell et al. (1997), brings us back to the FWFW questions, as in many cases 'stakeholder' is discursively employed to convey the feeling that all are being taken into account, while in fact this is not necessarily the case. When we observe how the word is actually employed in the EUCS and in the white paper, its referents are rather abstract—e.g. 'collaboration across borders and between different stakeholders can create valuable synergies' (SOG 2016, 10) or 'Cooperation with national cycling officers and stakeholders is a major component of the mission of such a focal point' (ECF 2017, 101)—so rather than the manipulative use observed in other kinds of texts (see Caimotto and Molino 2011) what we observe here is the attempt to conform to the dominant, economy-focused discourse, which, as explained above, can have a detrimental effect when promoting changes that are more important than the economy and go beyond its scope.

6.4 Reframing Strategies and Concluding Remarks

As Lakoff explains, many people have the wrong frames in their brain circuitry and that prevents them from understanding the environmental crisis. 'Those wrong frames don't go away. You can't just present the relevant facts and have everyone erase significant circuitry in their brains. Brains don't work that way. What is needed is a constant effort to build up the background frames needed to understand the crisis, while building up neural circuitry to inhibit the wrong frames. That is anything but a simple, short-term job to be done by a few words or slogans' (2010, 74). Lakoff repeatedly underlines the complexity of changing the frames, but also provides everyday helpful hints, underlining the importance of talking at the level of values and of finding general themes that incorporate the points we are making (79–80).

In order to become more effective when presenting all the relevant data that show how cycling can improve people's lives, we need to change the frames we employ when demanding more and better cycling infrastructure, stricter enforcement to grant higher levels of safety for people cycling, or the provision of all the kinds of services that will make life easier for people who choose to switch to cycling (e.g. secure parking spaces, changing rooms and showers in people's workplaces, easy intermodality with the possibility of taking one's bicycle on public transport). The first frame we need to counter is the conflict between road users, a lesson applied by an increasing number of advocates. There is no road war, people who drive a car are not enemies—sometimes they cycle themselves—and an increase in cycling is in the interest of all citizens. This is discursively achieved by talking about people, citizens or using the name of the inhabitants of the city being planned (e.g. Londoners) and presenting the advantages and positive effects as benefitting all people.

The second, more difficult, frame that needs to be changed is that of economic growth as the ultimate goal of policies and measures, what Mautner describes as the need to remain above 'the bottom line' that is increasingly becoming the social principle driving any human relationship. The argument 'we should promote cycling because it's (also) good for business' is weak, it reinforces the neoliberal, automentality-related framing that we are trying to counter. If the automobility system is good for business, why should anyone want to change it for another system, which—until implemented—is only *potentially* good for business? Of course, as cyclist advocates, we know the answer to this question: because it would make our cities safer, more liveable, more pleasant and socially cohesive, because it would reduce the number of friends and relatives (not 'people') who are likely to be hit and killed by a vehicle (not 'die in an accident') or likely to die of pollution-induced lung cancer because we would all feel happier and healthier. Aren't these three lines much more appealing to any human being than any statistical economic evidence, no matter how sound and well-documented? Wouldn't these arguments prove much more effective when employed by politicians to obtain consent? Cyclists advocates—whether individuals or large and established advocacy groups—need to beware of the risk that the frustration of not being listened leads to the increasing attempt to conform to the discourse of those who are not listening, entering a spiral of self-harming bad framing. And, even if this goes beyond the scope of this book, readers will recognise a similar kind of spiralling when it comes to much discourse promoting environmental protection and trying to counter climate change.

Reframing is a complex, demanding operation that requires a long time. Apart from the good amount of data which has already been published to prove the positive effects of increasing cycling levels, as Lakoff suggests, we need stories and narratives to exemplify values and rouse emotions and we need to address everyday concerns. The next chapter moves away from written, institutional discourse and investigates the transcripts of interviews carried out under the Aldred's *Cycling Cultures* project (2012), in order to retrieve the reasons why people cycle, what makes them happy, what are their difficulties and how these relate to possible solutions. These will then be employed in Chapter 8 to draft a guide to reframing and relexicalisation strategies.

References

Aldred, R. (2012). *Cycling Cultures: Summary of Key Findings and Recommendations.* https://westminsterresearch.westminster.ac.uk/item/8z5y3/cycling-cultures-summary-of-key-findings-and-recommendations. Accessed 2 June 2019.

Alexander, R. J. (2009). *Framing Discourse on the Environment: A Critical Discourse Approach.* London and New York: Routledge.

Anthony, L. (2019). *AntConc (Version 3.5.8) [Computer Software].* Tokyo, Japan: Waseda University. https://www.laurenceanthony.net/software.

Brown, L. R. (1981). *Building a Sustainable Society. A Worldwatch Institute Book.* New York: W.W. Norton.

Caimotto, M. C., & Molino, A. (2011). Anglicisms in Italian as Alerts to Greenwashing: A Case Study. *Critical Approaches to Discourse Analysis Across Disciplines (CADAAD), 5*(1), 1–16.

ECF. (2017). *EU Cycling Strategy: Recommendations for Delivering Green Growth*. https://ecf.com/eu_cycling_strategy. Accessed 20 May 2019.
EU. (2015). *Declaration on Cycling as a Climate Friendly Transport Mode*. http://www.eu2015lu.eu/en/actualites/communiques/2015/10/07-info-transports-declaration-velo/07-Info-Transport-Declaration-of-Luxembourg-on-Cycling-as-a-climate-friendly-Transport-Mode—2015-10-06.pdf. Accessed 10 June 2019.
EU. (2018). *Graz Declaration*. https://www.eu2018.at/latest-news/news/10-30-Graz-Declaration.html. Accessed 10 June 2019.
EU. (2019). *Cycling Policy and Background*. https://ec.europa.eu/transport/themes/urban/cycling/guidance-cycling-projects-eu/cycling-policy-and-background_en. Accessed 10 June 2019.
Fairclough, N. (2000). *New Labour, New Language?* London: Routledge.
Freeman, R. E. (1984). *Strategic Management: A Stakeholder Approach*. Boston, MA: Pitman.
Gonella, F., Almeida, C. M. V. B., Fiorentino, G., Handayani, K., Spano, F., Testoni, R., & Zucaro, A. (2019). Is Technology Optimism Justified? A Discussion Towards a Comprehensive Narrative. *Journal of Cleaner Production, 223*, 456–465.
Junker, J. (2014). *A New Start for Europe: My Agenda for Jobs, Growth, Fairness and Democratic Change*. https://ec.europa.eu/commission/sites/beta-political/files/juncker-political-guidelines-speech_en.pdf. Accessed 4 June 2019.
Kahlmeier, S., Racioppi, F., Cavill, N., Rutter, H., & Oja, P. (2010). "Health in All Policies" in Practice: Guidance and Tools to Quantifying the Health Effects of Cycling and Walking. *Journal of Physical Activity and Health, 7*(1), 120–125.
Keil, R. (2007). Sustaining Modernity, Modernizing Nature: The Environmental Crisis and the Survival of Capitalism. In R. Krueger & D. Gibbs (Eds.), *The Sustainable Development Paradox: Urban Political Economy in the United States and Europe*. New York and London: The Guilford Press.
Lakoff, G. (2010). Why It Matters How We Frame the Environment. *Environmental Communication, 4*(1), 70–81.
Lakoff, G., & Johnson, M. (2013). *Metaphors We Live By*. Chicago and London: The University of Chicago Press.
Machin, D., & Mayr, A. (2012). *How to Do Critical Discourse Analysis*. London: Sage.
Mautner, G. (2010). *Language and the Market Society: Critical Reflections on Discourse and Dominance*. London: Routledge.
Mayor of London. (2018). *Mayor's Transport Strategy*. Greater London Authority. https://www.london.gov.uk/what-we-do/transport/our-vision-transport/mayors-transport-strategy-2018. Accessed 18 July 2019.

Meadows, D. H., Meadows, D. L., Randers, J., & Behrens, W. W., III. (1972). *The Limits to Growth: A Report for the Club of Rome's Project on the Predicament of Mankind*. New York: Universe Books.

Mitchell, R., Agle, B. R., & Wood, D. J. (1997). Toward a Theory of Stakeholder Identification and Salience: Defining the Principle of Who and What Really Counts. *The Academy of Management Review, 22*(4), 853–886.

OED Online. (2019). Oxford University Press. www.oed.com.

Oosterhuis, H. (2019). Entrenched Habit or Fringe Mode: Comparing National Bicycle Policies, Cultures and Histories. In T. Männistö-Funk & T. Myllyntaus (Eds.), *Invisible Bicycle: Parallel Histories and Different Timelines*. Leiden: Brill.

Reagan, R. (1980, July 17). *Address Accepting the Presidential Nomination at the Republican National Convention in Detroit*. https://millercenter.org/the-presidency/presidential-speeches/july-17-1980-republican-national-convention. Accessed 10 August 2019.

SOG. (2016). *Sustainable Urban Transportation: Creating Green Liveable Cities*. https://stateofgreen.com/en/uploads/2016/06/Sustainable-Urban-Transportation.pdf?time=1565032648. Accessed 10 June 2019.

SOG. (2019). *Find Partners*. https://stateofgreen.com/en/partners/. Accessed 10 June 2019.

Stibbe, A. (2018). Positive Discourse Analysis: Rethinking Human Ecological Relationships in Fill. In A. Fill & H. Penz (Eds.), *The Routledge Handbook of Ecolinguistics* (pp. 165–178). New York and London: Routledge.

Urry, J. (2004). The 'System' of Automobility. *Theory, Culture & Society, 21*(4/5), 25–39.

Walks, A. (Ed.). (2015). *The Urban Political Economy and Ecology of Automobility: Driving Cities, Driving Inequality, Driving Politics*. London and New York: Routledge.

World Commission on Environment and Development (WCED). (1987). *Our Common Future*. Oxford University Press, Oxford. https://www.are.admin.ch/are/en/home/sustainable-development/international-cooperation/2030agenda/un-_-milestones-in-sustainable-development/1987–brundtland-report.html. Accessed 25 July 2019.

CHAPTER 7

The Words of People Who Cycle

Abstract This chapter investigates a corpus of interviews carried out for the ESRC project *Cycling Cultures in a Mass Motorised Society*. By looking at spoken discourse, this analysis retrieves what makes people choose cycling over other means of transport, as these arguments may prove effective in promotional material. The investigation focuses on road users labels, the use of metaphors, 'us vs. them' discourse, framing and reframing strategies. The role attributed to 'culture' is observed together with the notion of lifestyle practices, exploring the role played by identity-related issues. The chapter closes with some hints for motivational campaigns retrieved from the observation of themes touched by the interviewees, highlighting the positive effects of slow cycling and connecting these aspects to the ecosophy behind the book and the notions of good life and well-being.

Keywords Culture · Lifestyle practices · Slow cycling · Ecology · Good life

After having observed cycling-related discourse in newspapers and institutional documents, we now move to the words of people who cycle. By looking at spoken discourse, we aim to retrieve what makes people choose cycling over other means of transport, as these arguments may prove effective along with, or instead of, the rational arguments often

M. C. Caimotto, *Discourses of Cycling, Road Users and Sustainability*, Postdisciplinary Studies in Discourse,
https://doi.org/10.1007/978-3-030-44026-8_7

employed in promotional material. We shall also find whether traces of marketised discourse or 'us vs. them' instances are present in spoken interviews as well, paying attention to the discourse strategies employed. If we aim to activate frames that foster our ecosophy, one of the most straightforward ways is observing the language of people who have already developed those frames, that is people who favour cycling for a wide variety of reasons beyond the economic advantages highlighted by much institutional discourse. Of course, people can also reproduce and reinforce the problematic discourses that we have observed in previous chapters, the goal of this investigation is to retrieve useful hints to construct new framing.

7.1 The Cycling Cultures Project and Our Approach

The interviews employed for this investigation are part of an ESRC publicly available database, which was created as part of the *Cycling Cultures in a Mass Motorised Society* project (Aldred 2012b). The project ran from January 2010 to December 2011 and examined experiences of cycling in four relatively high-cycling English urban areas, Cambridge, Bristol, Hackney (Inner London) and Hull. The people interviewed are cyclists and, Aldred explains, the project 'provides an insight into cycling in English urban areas among people who are not mostly "hardcore" or "proper cyclists", and who cycle in areas where cycling is *relatively* normalised' (Aldred 2012a, 4 emphasis in the original). The aim of the project was to observe what made cycling become more normalised in those areas (in 2001 all of the four areas had a cycle to work rates at least twice as high as the UK average) compared to the national situation, described as 'unfavourable'.

As Aldred explains, three-quarters of the interviews were carried out with everyday cyclists, while the remaining quarter consisted in 'stakeholders', i.e. people working for the local city council, people involved in cycling-related businesses—such as selling bicycles or running a cycle courier firm—and people involved in cycle campaigning in various forms. The pilot corpus was collected in Cambridge in 2008 and the interviews were semi-structured, while the interviews with everyday cyclists collected in 2010–2011 were relatively unstructured. The interviewers began by asking 'can you tell me about cycling in relation to your life' and the following questions were related to the interviewee's response.

The areas of interest often emerged spontaneously and covered the cycling habits of friends and family, one's self-perception as a 'cyclist', experienced theft or crashes, opinions about the local cycling environment and infrastructure. Interviews with stakeholders usually opened with a description of their role in relation to cycling and the following questions were partly pre-planned and tailored and partly based on the stakeholder's response (Aldred 2015, 694).

The corpus consists in 1,819,726 tokens, divided into 9 subcorpora based on the geographical location and the category of interviewees, i.e. five subcorpora for everyday cyclists—the Cambridge Pilot interviews from 2008 did not include stakeholders—and 4 subcorpora for stakeholders. The Everyday Cyclists (EC) subcorpus consists of 1,378,530 tokens, and the Stakeholders (SH) of 441,196. The investigation of the corpus was carried out at the same time through a qualitative and a quantitative approach, in order to access the data from different perspectives. This means that the activity of reading interviews in full was alternated with a 'vertical' reading following some statistically relevant words or expressions, thus aiming to retrieve discourse patterns to be employed for reframing while reducing the influence of my own perspective.[1] The corpus has been investigated with Lancsbox (Brezina et al. 2018).

7.2 'Cyclists' and Other Road Users

Throughout the previous chapters we have observed the presence, reification or absence of categorisation of road users under specific labels such as 'cyclist', 'pedestrian', 'driver', 'motorist'. We shall thus start our corpus-aided investigation by observing the noun cyclist(s) and its collocates. In the EC subcorpus we find 2290 occurrences of the lemma, and when we observe its collocates (CPN: 03-MI (4.0)/L5-R5/C: 15.0-NC: 15.0) we see that other road user categories are prominent, the most statistically significant being 'motorist' (7.35) followed by 'pedestrian' (7.09) and 'driver' (5.85) (Fig. 7.1).

As we may expect, some instances display the typical 'us vs. them' structure:

[1] Rachel Aldred, The University of East London, the Economic and Social Research Council and the UK Data Archive bear no responsibility for the further analysis and interpretation of the data presented in this book.

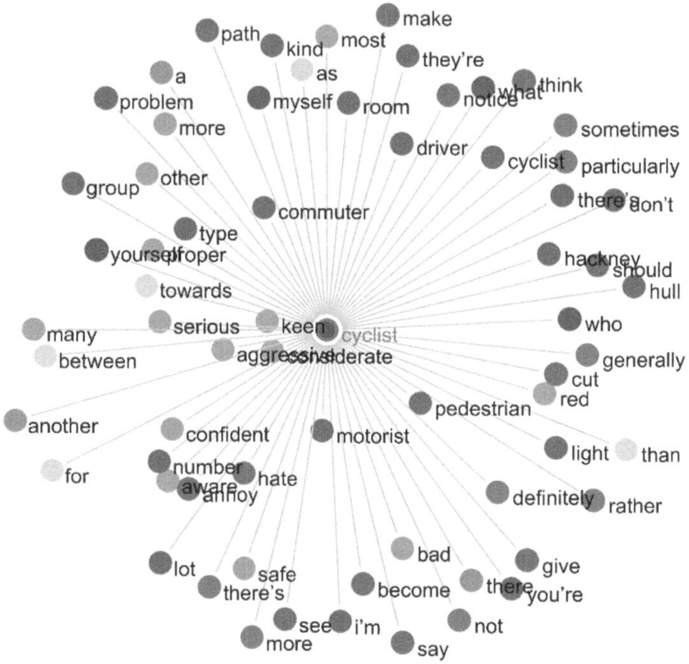

Fig. 7.1 Collocation network for cyclist, colours indicating word classes

> But there's a group of motorists who psychopathically have a hatred of cyclists for some reason. They think we're fair game. A lot of them are black cab drivers I've got to say. And I think there's nothing more they enjoy than leaving you four inches of space as they go past you at 50 miles an hour. It's like a sport. And I get… I have got pretty angry. (Hackney EC)

In the first sentence, we notice that the speaker uses state verbs (be, have) and presents his opinion as a fact. In order to tone down his sweeping judgement, he uses the face-saving expression 'a *group* of motorists', implying that not all motorists are bad. As the speaker gets more emotional, he shifts to 'they' and 'we', introducing the metaphor of 'fair game'. While nowadays this metaphor is mainly used figuratively to refer to someone or something that people are allowed to criticise, the speaker rather seems to draw on the meaning of 'game' defined as 'figurative and in extended use. Any object of pursuit; (also) an object in view, an aim, a goal. Often with conscious hunting metaphor' (OED Online 2019). At the end of this passage, the interviewee explicitly refers to his anger and we can imagine that the situation described is one that he has experienced personally. He is turning it into a universal experience by using 'you', the simple present tense and the state verb to be, together with the superlative expression 'nothing more they enjoy'. The way he uses the metaphor of fair game is revealing of his feelings.

This instance is representative of the kind of conflictual discourse that can arise from the moments of danger that people who cycle on roads shared with cars are likely to have experienced. Feelings of fear and helplessness can thus be expressed into sweeping statements against all people who drive, using the typical discursive patterns that characterise discriminatory discourse triggered by fear. This is linked to what we have observed in Chapter 4: while newspapers may reproduce the discursive patterns of elite racism as described by van Dijk (1992), everyday conversations may reproduce discursive patterns of overgeneralisation and face-saving strategies. Searching for collocates of the lemma 'motorist' in the EC subcorpus (CPN: 03-MI (3.0)/L5-R5/C: 15.0-NC: 15.0), we obtain 12 occurrences of 'motorist(s)' and 'they' occurring together, 'they' being the fifth significant collocation (4.31):

Corpus: Everyday Cyclists| Search Term: {motorist}| Occurrences: 12/126 (0.09)| Texts: 5| Filter applied|

Left	*Node*	*Right*
like it goes viral, when erm... young	**motorists,**	male motorists, **they** would, **they** would pull
goes viral, when erm... young motorists, male	**motorists,**	**they** would, **they** would pull up alongside
daughter reckons that **they** should get all	**motorists**	to do a bicycle... You know, the
from bicycles, so even if **they,** as	**motorists,**	regard cyclists as a menace, **they** can
find them useful, but I think other	**motorists**	hate them, don't **they.** Taxi drivers hate
to abuse the crossings which irritates the	**motorists,**	**they** hate us even more—vicious circle.
sometimes it's really quite funny when clearly	**motorists**	feel **they** have to overtake you even
and there is definitely a sense amongst	**motorists**	that **they** would rather cyclists just didn't
good because I think **they** don't annoy	**motorists**	in the same way as speed cameras
you, makes you more aware of what	**motorists**	don't do. The way **they** don't look,
oh. Yes they, **they** irritate me, the	**motorists**	(laughs) when I want to be home
I think sometimes it's down to the	**motorist**	as well, **they** don't look, it's the

As mentioned in previous chapters, the divisive attitude that labels people as separate groups is not constructive and it is important to observe its dynamics in order to avoid reinforcing the conflict. As we observed in the Mayor of London's Strategy (2018), a good way of avoiding conflict consists in focusing on people rather than road user categories. In fact, in the corpus we also find instances in which the interviewee discusses and challenges the use of road user labels, as it can be detrimental, removing the humanity of the people being discussed:

> So I rather resist the use of the word "cyclist" because when you have cyclist campaigns and cyclists interests, you know, it sort of labels people and, and it, but, and it sounds pedantic but if you actually talk in terms of people on bicycles, people on cycles the, the, the tenor of the sentence changes, you know what I mean? [...] if you just label people they become two dimensional and sometimes you get almost a sort of warfare between the motorist and the cyclist, whereas usually it's the same person, just a difference in the number of wheels. (Cambridge Pilot)

Another interviewee problematises the implications other people will infer if she uses the label 'cyclist' to describe herself:

> I wasn't sure where to draw the line or what fitted the mould, so to people that drive I'm like "Yeah, I'm a cyclist", but to cyclists I am just "Yeah, I cycle to work" (laughs) So, it feels weird, and in a way, I don't know why and I don't know why I attach so much importance to the label and it depends who I am with as to what label I use or how I describe it, and sometimes I feel like I am boasting to people when I am like, if I say I am cycling to work, as if I am really green and really aware of the environment and really like, you know, coming from that angle, so I don't want to seem too pretentious... that is probably not the right word, but I don't want to seem like I'm someone I'm not, if I'm like "Yeah, I cycle to work" and saying I am better than you and I am trying to like do my bit for the environment, when really, in all honesty, it's because of costs that I do it, it's not really... an added benefit is that, yeah, I am not in a car, erm... but... (Cambridge EC)

In this passage, the speaker's self-perception of her identity as a cyclist is expressed through a number of metaphors 'draw the line', 'fit the mould', 'label' and 'from that angle'. This high use of metaphors does not seem to be a feature of this interviewee's style in general, so all the metaphors employed here to reveal a sense of discomfort with the issue at hand. This feeling can be ascribed to the use of labels in general, but also to the reified image of the cyclist as someone 'really green and really aware of the environment', an identity with which the interviewee does not feel at ease. Even more significant for our purposes is the passage in which she equates saying 'I cycle to work' to saying 'I am better than you': this shows how framing cyclists as morally superior to other people can generate identities with which even people who actually behave in environmentally friendly ways have trouble.

Moreover, it is worth observing how doing something for the environment and doing something in order to save money are discursively presented as in contrast, as the speaker appears to frame environmental reasons as moral and cost-related reasons as material. The repetition of 'really' and the use of 'in all honesty' discursively imply that only people who do not care about the costs can claim that they cycle for environmental reasons and that the latter are mostly ideological rather than material. The reduction in pollution a person generates by choosing to cycle is thus framed as irrelevant on a material level and only significant from a moral perspective. This is in line with Lakoff's (2010, 77) observations concerning the absence of established frames in public discourse to see the economic and ecological meltdown as consequences of the same causes, while there are established frames conceiving them as opposites.

Road user labelling is also employed by other interviewees who discuss their identities as both a motorist and a cyclist, explaining how this affects the way they perceive other road users.

> I always think being a motorist makes you a lot better cyclist. [...] you think about things like looking round before you pull out to overtake a parked car which if you didn't drive you probably wouldn't realise why that mattered to cars so. (Hull EC)

We notice how the speaker opens with a first-person pronoun and within the same sentence shifts to a general second person. While he is speaking about his own experience and self-perception, he transforms the statement into a generalisation concerning people who cycle and drive. It is worth noticing that, as he is sketching an identity of a cyclist–motorist, he uses a metonymy[2] and refers to other motorists as 'cars'. This is, of

[2]A figure of speech characterised by the action of substituting for a word or phrase denoting an object, action, institution, etc., a word or phrase denoting a property or something associated with it, e.g. as when referring to the monarchy as 'the crown' or the theatre as 'the stage'; an instance of this (OED Online 2019).

course, a widespread and common metonymy and entails several relevant issues, as the attribution of human actions to cars sometimes is a way to reduce or background the driver's responsibility and, more in general, identifying cars as human subjects contributes to the feeling of being personally targeted and contained when traffic-calming policies are introduced (see Goddard et al. 2019).

Conflict also emerges in the lexical choices employed to talk about lanes. Among the most significant collocates of the lemma 'lane' in the EC subcorpus (CPN: 03-MI (3.0)/L5-R5/C: 5.0-NC: 5.0) we find 'segregated' (9.58) and 'dedicated' (9.01), while in the SH subcorpus (CPN: 03-MI (3.0)/L5-R5/C: 5.0-NC: 5.0) the focus is rather on the technical features of lanes: metre (9.32), carriageway (8.58), dual (8.44), wide (8.34) are the first significant collocations. The SH subcorpus reveals another significant lexical choice in the phrase 'traffic lanes' which occurs 11 times, while it is never employed in the EC subcorpus. While the adjective 'segregated' can trigger negative associations and 'dedicated' appears preferable, the fact that we talk about lanes that are exclusively for buses or bicycles discursively implies, and reinforces, that the norm on the road is cars and motor vehicles. Using the phrase 'traffic lanes' shifts the perspective and also adds the negative prosody attached to the word 'traffic'.

7.3 Culture, Identity and Lifestyle

Inspired by Cox's work on 'cycling cultures' (2015), a search for 'culture (s)' in the corpus revealed 157 occurrences. One-third of these (50 occurrences) are 'cycling cultures' and when observing the context, most interviewees describe a cycling culture positively, either appreciating the local cycling culture—mainly in Cambridge and Hackney—or, in fewer cases, admiring the cycling culture of another place and wishing the cycling culture of the place where they live could be enhanced and made stronger. Research concerning race-related discrimination can prove helpful once again to investigate this prominence of culture, in the same way it was employed as a source of inspiration in Chapter 4. More specifically, the idea that specific 'cycling cultures' or 'car cultures' exist is reminiscent of debates around the notions of 'cultural essentialism' or 'cultural fundamentalism' as new forms of racism (see Siebers and Dennissen 2015; Grillo 2003). The idea of 'new racism' consists in theorising that—because nowadays discourses of biological superiority are considered unacceptable across society as a whole—hostility and discrimination against immigrants and minorities in recent decades has moved to envisage incompatible cultural differences, grounding the incompatibility in people's culture rather than in people's blood and skin colour.

As far as our investigation is concerned, the ways in which the interviewees employ the notion of 'culture' might be part of a much broader tendency towards what Hannerz (1999) has labelled 'culturespeak'. Of course, the occurrences of 'culture (s)' in the interviews are partly influenced by the fact that Aldred's research project was titled *Cycling Cultures*, by which the authors referred to the tendencies that made people living in the four selected locations cycle more than elsewhere. But in some of the interviews, the notions of 'cycling culture' and 'car culture' appear much more fixed and closer to the notions of 'cultural essentialism' mentioned in the debates around contemporary discrimination. The use of 'culture' in the project describes tendencies that are influenced by various factors and can be steered in other directions. Also Cox (2015, 38) highlights ways in which the notion of 'culture' can be employed to bring together diverse cycling cultures and better understand bicycle politics, resisting marginalisation—which he compares to the marginalisation of women.

Still, if we look at the evidence from the corpus, the way in which 'culture' is envisaged by some interviewees is something much more fixed and tied to a geographical place, an intrinsic local feature that will hardly be influenced by any policy. The following quotations are a selection from the 108 occurrences of 'culture (s)' out of the 157 in EC subcorpus that within a 20-word span context to each side of the node have the words 'cycling' or 'bike' or 'car'.

> Cycling in relation to my life? Well, it wasn't really part of my life until I moved to Cambridge. Erm… my husband always cycled in London. Lived in London for fifteen years and he always cycled to work every day but I never did, and erm… as soon as we moved to Cambridge, obviously the culture of cycling in Cambridge is so, you know, it's flat, everyone does it. Bikes are easy to get cheap, you know, you can get around so easily in a small city like Cambridge. So, I sort of started cycling from there really, when we came to live in Cambridge. (Cambridge EC)

> So erm… but yeah, I presume drivers don't see me and I think that's the best way, in, there is just, you know, there's no cycling culture. They don't, car drivers presume that cyclists are a pain in the arse, the roads are for them, and as long as you have that mentality and, you know, that as a cyclist, you're much more likely to be safe so… (Bristol EC)

> Erm. It's just a completely different....culture do you know for... er cycling. Most people in Hull like... erm... they'll go and start up with a bike when they're younger now, it's school age but then it's, it's just that much easier do you know for, for car, it's sort of thing but over in Holland it 's.. just more geared to cycling. (noises). Erm... it, er, because all the roads are set up.. erm.. in and well er it's a flat country so... it's just more of a culture thing in Holland. (Hull EC)

> so there is a car culture and again it comes from the car lobby... the car industry who really are blocking any developments to more sustained driving so... they feed that just like they feed the pro-nuclear lobby and you know there is very strong anti-windmill thing there and it is not a coincident it is just manipulated by them out there so erm... it is the same as cars. (Bristol EC)

Interviewees appear to discuss 'culture' as something that belongs to a place, either because the land is flat and makes cycling easier or because the road infrastructures favour a 'car culture' or a 'cycling culture'. The last speaker describes 'car culture' as promoted by external forces, but also in this case the chances that change could be brought about are implied to be non-existent.

Moving away from the specific discourse of cycling and embracing a broader perspective, Hannerz's work (1999) envisages the fixedness of 'culture' as a general tendency of contemporary public discourse, especially employed to explain the origins of situations of conflict. Hannerz highlights the contrast existing between an academic notion of 'culture'—which embraces multiculturalism as positive—and the notion of cultural fundamentalism 'which may perhaps be celebrationist *within* a culture, but not across "cultural borders"' (Hannerz 1999, 399 emphasis in the original). The author convincingly argues that scholars involved in cultural studies should attempt to promote a response revolving around the notion of 'process', fighting the tendency of cultural fundamentalism 'to naturalize cultural immutability and persistence' and bringing the relevance of reproduction and change back into the equation. He explains 'it would seem helpful to make more continuously visible how both persistence and change in culture depend on human activity; and how in contemporary, complex social life, the combined cultural process, and the overall habitat of meanings and practise in which we dwell, is the outcome of the variously deliberate pursuit by a variety of actors of their own agendas, with different power and different social and spatial reach, and with foreseen or unanticipated consequences' (401).

The view of culture as something given and immutable may at first appear in contrast with what has been described as 'lifestyle practices', that is, as Machin and van Leeuwen explain (2005, 582), ways in which people signify their individual identities through appearances, gradually moving away from the more stable signifiers that would communicate one's social position in terms of class, gender and age, social relations and social roles we play within socially regulated activities. Lifestyle choices allow people to announce, through their appearance, their 'interpretations' of the world, their affiliation with certain values and attitudes (584) and this is mainly communicated through consumption patterns and attitudes. The contrast is thus only apparent, and in fact, there is a deep connection between the tendency to express one's ideas and beliefs by buying certain products and services and the tendency to see one's individual actions as insignificant in terms of their ability to bring some kind of change in society. Coming back to the interviews, some of the responses reveal a weakening perception of the connection between what I do as an individual and how this is likely to affect the local culture, as if culture was perceived as something static and, therefore, cycling is possible only where a cycling culture already exists.

A society in which the values and beliefs of individuals are expressed mainly through the purchasing of goods and services is a society in which beliefs and values cannot be carried forward outside the categories that marketing experts layout to label existing and potential customers. Any attempt to promote change outside this setting is likely to be attacked and disqualified as hypocritical, especially when the change being promoted aims to question the current system of consumption in order to envisage and enact new ways of living that can prevent the worst outcomes of the effects of climate change. Within this framework, promoting everyday cycling as part of one's lifestyle—a projection of one's environmental consciousness through the purchase of cycling gear—is a dangerous path to follow. In the short term, it undoubtedly has the advantage of drawing on existing and effective marketing strategies that will fit people's frames and, if well planned, may prove successful in convincing a number of people to shift to a bicycle as their main means of transport. Nevertheless, in the medium-long term, it may trap a mobility choice that actually is good in many ways into just another fashionable consumer's choice that simply reinforces the dominant framework of marketised options, from which any belief and value is diluted into the mere projection of individual identities unable to affect the society around them. Our next step, thus, is to dig into the words of the interviewees to retrieve discourse strategies that may be employed not only to promote everyday cycling but also to contribute to the deconstruction of a marketised, neoliberal and consumeristic mental framework.

7.4 Hints for Motivational Campaigns

What emerges from the interviews is that the main reasons that make people want to cycle are linked to their well-being: cycling on a regular basis makes people feel happier. This increase in happiness can be attributed to specific changes in one's life, for example losing excessive weight, falling ill less frequently, obtaining relaxation from time spent outdoors—away from electronic devices and other sources of stress—engaging more with other people. Of course, different people will obtain more benefits from one or two specific changes and ignore others that are less significant for them. While on first impulse most of us would probably list the effects by separating those benefitting one's mind from those benefitting one's body, it is worth reframing and underlining how the two things are connected, a link best described by the following speaker:

> .... What do I like and love about cycling?.......... Erm... What do I like and love about cycling?.... Erm... The way it can connect you with, with .... being alive through.. the weather... through... the blood pumping round your body.. through....the social aspect of... talking to people on your bike and working with people to fix problems..... it's....it's a wonderful... metaphor for life (laughs). .... (laughs).. (Hull, EC)

The explicit reference to metaphors—the expression of an abstract concept through an implicit comparison with something concrete—reminds us of the strength of the connection between the world of ideas and the physical world that we tend to conceive of as separate. Practising cycling and promoting cycling are effective ways of becoming increasingly aware of the relevance of well-being for both mind and body and their interrelation, especially for people who live in urban environments, practise little or no sports and are much more involved in abstract rather than concrete everyday activities. The tendency to keep body and mind separate is discursively reflected in institutional talk when 'utilitarian cycling' is conceived as separate from 'leisure cycling' and when the positive effects for people's bodies are presented as separate from the positive effects in terms of stress reduction. According to Aldred (2012a, 29), replacing 'utility cycling' with 'everyday cycling' acknowledges that motivations and purposes often overlap and focuses more on leisure and socialisation.

Popan (2019) highlights the positive aspects of slow cycling, which is more natural for the human body, reawakens bodily senses that are otherwise numbed inside automobiles, allows appreciation of the urban environment and 'enables playful interactions with other cyclists, conversations, as well as the transformations of the utilitarian space of the road into a more socially open space' (89). Popan significantly argues that investigating slower mobilities is essential if we want 'to propose alternatives to contemporary growth-oriented capitalist societies' and that 'a slow bicycle system can contribute to establishing a norm of sufficiency (Gorz 2010) which is essential to avoid both economic and ecological catastrophes' (Popan 2019, 89).

The positive effects of cycling on one's body are framed by various interviewees as a way for them to avoid having to think about exercising and doing things like going to a gym. Of course, they are exercising and improving their health, but one of the reasons why they like everyday cycling is that it allows them to do these things without thinking of themselves as actually leaving the house in order to exercise. These aspects could prove relevant when designing promotional campaigns. For example, suggesting that cycling would result in weight loss is likely to activate negative associations of sacrifice, failure and fatigue in the mind of people who have always struggled with weight loss—that is, exactly those people who would benefit the most from an increase in their levels of physical activity. The following relevant passages from the interviews show that occasionally eating some extra, not necessarily healthy, food without putting on weight as they usually would is perceived as a source of motivation for some interviewees:

> I've always said lightly "Oh I cycle so I can eat cake", you know. Erm because I'm very keen that one shouldn't sound as if you're, you know, being judgemental but there is a very genuine health thing, you know, not so much the eating of the cake but then certainly the using of the energy and the cardio vascular whatsits so I suppose, because I've always been fairly health conscious, even before I trained as a nurse, that's probably one of the things erm. (Cambridge Pilot)

> "We need to lose weight" or even when I used to be at work they used to say, "Oh you can eat a cream bun can't you?" do you know. And they used to be really envious and then they'd turn round and say, "Yeah but you cycle" and I'd say, "Well why don't you?" (Hull EC)

> there's a brilliant cake shop in Monmouth. Absolutely superb. They do these brilliant hot chocolates where they serve you this big glass of hot milk and then a little tub of chocolate buttons, like proper chocolate and you stir it in. It's like, oh, it's heaven. Worth the trip (laughs). [...] Then you can, yeah, absolutely. Eat as much as you like (laughs). (Bristol EC)

It is important though, when designing campaigns, to bear in mind the potential pitfalls. Rather than conveying the message 'cycle so you can eat cake', reframing could focus on the way in which cycling—or more, in general, practising some physical activity as part of one's everyday habits—allows people to improve their perception of their bodies and of how much calories they actually need to feel well. While not promoting sugar intake, reframing could focus on the fact that not being overweight is a source of well-being, as much as knowing how to respond to the feeling of hunger generated by physical activity, for example by eating dried fruit or nuts. At the same time, such reframing would deconstruct the efficiency frame—conveyed by the idea that cycling saves the time one would spend in the gym—and the idea that we need good-looking bodies in order to be happy. The focus would shift on well-being, in terms of health, in terms of the bodily pleasure provided by cycling and that of knowing how to balance the intake and burning of calories.

Promoting the advantages in terms of individual well-being can be an effective strategy, but it still has the downside of focusing on individualism. The observation of N-grams in the EC subcorpus revealed features that can be employed to create a different narrative. The most frequent trigram is 'a lot of' (1514 occurences). Its most frequent collocate to the right is 'people' (240). In most cases, 'a lot of people' is employed by the speaker to make judgements about the general situation in relation to cycling, these can be either positive—e.g. 'it's small measures like that that have actually, you know, enabled a lot of people to cycle, so it's, it's small key promotional efforts' (Cambridge pilot); 'You get a lot of people out in the summer, actually. You notice them there on a sunny day, how many more cyclists are out'. (Bristol EC)—or judgements about what other people do wrong when they have little cycling experience—e.g. 'And where your seat should be, you see a lot of people with it too low or too high' (Hackney EC). While this confirms Aldred's (2013) observations about the stigmatisation of cyclists who are not competent, it also reveals that people who cycle tend to care about the people around them, about what they do in terms of travel choices and they pay attention to how people who cycle could be helped to improve their cycling experience.

This is worth taking into account when designing motivational campaigns. Rather than creating messages aiming to convince people to cycle more, to start cycling or to respect cyclists when driving—all messages discursively implying that the number of people who cycle is (too) small—the intention of switching to cycling could be taken for granted, and discursively framed as the obvious choice, while the actual message could consist in a piece of advice helping people who are new to cycling. It is of utmost importance that the campaign should be designed as addressing only people with little cycling experience, otherwise people who cycle regularly could perceive it as patronising, and the whole project could backfire. In particular, care should be taken against the risk of creating the discourse pattern usually described as 'blaming the victim'. While it is wise to underline the importance of wearing clothes that make you visible when riding, or to recommend that it is necessary to lock one's bicycle properly, everyday cyclists—who already feel regularly criticised and blamed—could react by highlighting that the actual problem lies in the lack of attention from people who drive or the low levels of safety and the proliferation of theft in general. Framing the advice as addressing only people who want to take up cycling, rather than any cyclist, could redress the balance, presenting the messages as the kind of advice that people with more experience would give to someone new to cycling. The campaign could thus have a general catching question addressing the recipients—e.g. 'Want to take up cycling?'—and then various different messages providing tips for inexperienced cyclists and highlighting the advantages that everyday cycling would bring along. Depending on the medium employed, the advice could be provided by an experienced cycle user who cares about fellow humans with whom people who cycle regularly could identify.

7.5 Concluding Remarks

The corpus-aided observation of spoken discourse reveals some problematic aspects, which need to be taken into account when designing promotional material, and some positive discourse features that can be employed to create new narratives. Referring back to the ecosophy we outlined in the introduction, the discourse features analysed above contribute to the creation of 'beneficial discourses which actively identify the "good life" with something other than consumerism' (Stibbe 2014, 125). The interviews confirm the need to reduce conflictual discourse, to reframe the perceived contrast between economy and ecology and to reinforce the idea that cultures are a process in constant change. At the same time interviewees reveal that the promotion cycling can be employed as channel through which new framing can be constructed. This new 'good life', rather than employing neoliberal self-styling strategies, revolves around a more human-focused, community-oriented vision of society. Empathy—towards fellow citizens, other living creatures and the ecosystem—is pivotal and well-being derives from improved health, benefiting the mind and the body at the same time, and improved feelings of safety and harmony with the system of which we are part.

References

Aldred, R. (2012a). *Cycling Cultures: Summary of Key Findings and Recommendations.* https://westminsterresearch.westminster.ac.uk/item/8z5y3/cycling-cultures-summary-of-key-findings-and-recommendations. Accessed 2 June 2019.

Aldred, R. (2012b). *Cycling Cultures in a Mass Motorised Society* [computer file]. Colchester, Essex: UK Data Archive [distributor], September 2014. SN: 7545 http://dx.doi.org/10.5255/UKDA-SN-7545-1.

Aldred, R. (2013). Incompetent or Too Competent? Negotiating Everyday Cycling Identities in a Motor Dominated Society. *Mobilities, 8*(2), 252–271.

Aldred, R. (2015). A Matter of Utility? Rationalising Cycling, Cycling Rationalities. *Mobilities, 10*(5), 686–705.

Brezina, V., Timperley, M., & McEnery, T. (2018). #LancsBox v. 4.5 [software]. Available at http://corpora.lancs.ac.uk/lancsbox.

Cox, P. (Ed.). (2015). *Cycling Cultures.* Chester: University of Chester Press.

Goddard, T., Ralph, K., Thigpen, C. G., & Iacobucci, E. (2019). Does News Coverage of Traffic Crashes Affect Perceived Blame and Preferred Solutions? Evidence from an experiment *Transportation Research Interdisciplinary Perspectives.*

Gorz, A. (2010). *Ecologica.* Chicago, IL: University of Chicago Press.

Grillo, R. D. (2003). Cultural Essentialism and Cultural Anxiety. *Anthropological Theory, 3*(2), 157–173.

Hannerz, U. (1999). Reflections on Varieties of Culturespeak. *European Journal of Cultural Studies, 2*(3), 393–407.

Lakoff, G. (2010). Why It Matters How We Frame the Environment. *Environmental Communication, 4*(1), 70–81.

Machin, D., & van Leeuwen, T. (2005). Language Style and Lifestyle: The Case of a Global Magazine. *Media, Culture and Society, 27*(4), 577–600.

Mayor of London. (2018). *Mayor's Transport Strategy.* Greater London Authority. https://www.london.gov.uk/what-we-do/transport/our-vision-transport/mayors-transport-strategy-2018. Accessed 18 July 2019.

OED Online. (2019). Oxford University Press. www.oed.com.

Popan, C. (2019). *Bicycle Utopias: Imagining Fast and Slow Cycling Futures.* London and New York: Routledge.

Siebers, H., & Dennissen, M. (2015). Is It Cultural Racism? Discursive Exclusion and Oppression of Migrants in the Netherlands. *Current Sociology, 63*(3), 470–489.

Stibbe, A. (2014). An Ecolinguistic Approach to Critical Discourse Studies. *Critical Discourse Studies, 11*(1), 117–128.

van Dijk, T. A. (1992). Discourse and the Denial of Racism. *Discourse and Society, 3*(1), 87–118.

CHAPTER 8

Positive Discourse Strategies and Concluding Remarks

Abstract The concluding chapter shows how the data presented in the book are inscribed in the wider discourse of the climate change crisis and how promoting more cycling can be one of the starting points from which effective reframing can take place. Employing Sheller (*Mobility Justice: The Politics of Movement in an Age of Extremes.* Verso Books, London and New York, 2018) and Halliday (*The Ecolinguistics Reader: Language, Ecology and Environment.* Continuum, London and New York, 175–202, 2001) to highlight the interconnection between Mobilities and Ecolinguistics, the chapter argues for reframing strategies that can further environmental protection and social justice at the same time. The positive discourse strategies retrieved throughout the book are brought together, focusing on preferred lexical choices. The positive effects of increased cycling are outlined and inscribed within Poli's 'revolutionary approach' (2011), in order to achieve the reappropriation of time-space and achieve the norm of sufficiency (Popan in *Bicycle Utopias: Imagining Fast and Slow Cycling Futures.* Routledge, London and New York, 2019). From this perspective, cycling becomes a pivotal point from which a new perspective can be framed in order to foster alternative worldviews.

Keywords Mobilities · Sufficiency · Revolutionary approach · Social justice · Environmental protection

M. C. Caimotto, *Discourses of Cycling, Road Users and Sustainability*, Postdisciplinary Studies in Discourse,
https://doi.org/10.1007/978-3-030-44026-8_8

In his concluding chapter for the *Handbook of Neoliberalism*, Purcell (2016) points out that as scholars we have become 'sick' as we are obsessed with neoliberalism. After Nietzche, he calls this sickness '*ressentiment*' and states

> as theorists we can only sing in the key of critique. We meticulously record and discuss the crimes and contradictions of neoliberalism. When we imagine the world we want instead, we can only speak in terms of not-neoliberalism, of cancelling out the current political-economic regime. When we act, we can only act in the register of protest, resistance, contestation, and refusal – of struggle *against* neoliberalism. [...] But we don't have to be sick. There is another way to think, another way to be. We can become obsessed with ourselves instead. What do we want to create? What are we capable of producing? Who are we capable of becoming together? What worlds, what ways of life have we already started to build, and how can we help them to grow, spread and flourish? (613)

Throughout this book, we have encountered several examples of discourses in which neoliberalism played a part in hindering the potential for cycling to grow, spread and flourish and they have been recorded and discussed, and we are probably guilty of joining the key of critique against which Purcell warns us.

A discourse analyst knows that if you negate a frame you reinforce a frame, still it was necessary to bring to light contradictions in discourse that are not necessarily obvious for the actors involved. With Plehwe (2016, 70) I believe that 'corporate social responsibility in combination with nudging of market citizens (consumers) provide for an alternative neoliberal frame that has yet to be challenged in a comprehensive and decisive way'. We thus need to be wary of some of the ways in which we promote sustainable mobility and cycling in particular. But if the previous chapters are guilty of *ressentiment*, this concluding chapter will compensate and focus exclusively on the way forward.

The aim here is to bring together all the positive strategies that have been observed throughout the book, to sketch new framing strategies to promote cycling and, by doing this, also provide useful insights that can be employed to frame the environment differently, as argued by Lakoff (2010), and in line with the ecosophy outlined in Chapter 1. In Sheller's words, 'the societal challenge consists in democratically transitioning toward both sustainable and socially just mobility systems in the face of looming threats such as global climate change, increasing natural disasters, and resource shortages. Sustainable mobility and transportation justice together are crucial to these transitions in mobility, and the two problems are fundamentally interconnected. But we must also envision this as a mobility transformation that goes far beyond automobility and involves far more than questions of transportation' (2018, 74).

With this aim in mind, we shall envision these changes from a systems thinking perspective, aiming to contribute to interventions in the system at the highest level, that of transcending paradigms (see Chapter 1, Meadows 2008). Our specific focus offers a reflection on how the SAE (Standard Average European) languages we speak are shaped by our society and how promoting changes in the language can contribute to bring change forward. Sheller (2018) underlines the interconnection between the protection of the environment and the actualising of social justice and this reflects Halliday's point when he states: 'the linguistic evidence suggests that these two transformations will have to take place together: that the hegemony arrogated by the human species is inseparable from the hegemony usurped by one human group over another, and that neither will come to an end as long as the other still prevails' (2001, 198–199).

What is the support that linguistics and discourse analysis can offer to achieve this goal? A good starting point is Lakoff's recommendation to 'go on offence, never defence': do not accept the frames you are trying to counter, do not negate them, do not repeat them, do not structure your arguments to counter them. Doing this will activate the wrong frames in the brains of the recipients and that will reinforce the frames you are trying to fight. Next, tell stories to exemplify the values and rouse emotions and find general themes or narratives that incorporate the points you are making. And finally address everyday concerns, be aware of the context, the visuals, the messenger, the body language. Use words that people can understand (Lakoff 2010, 79–80). When applying Lakoff's recommendations to cycling promotion, we need to be aware of the frames we are trying to counter and recognise the deep cultural and social implications of automobility. We then need to create and reinforce themes and narratives that move away from growth-oriented capitalist societies. A pivotal point is the heightened awareness of life experienced by people who cycle, an awareness which results from the bodily experience of moving and the reawakening of the bodily senses, a form of urban movement which, unlike most other forms, 'rescues us from a complete disengagement with the world' (Popan 2019, 138).

Through this perspective, we find a way to overcome the conflict between people who cycle and people who drive. The former is not better than other people, it is the activity of cycling itself which is better for any person in comparison to other forms of passive transport, as it favours health, social interaction and social justice rather than hindering them. Avoiding the word 'cyclists' allows to avoid the negative prosody it carries. As argued by Popan (2019, 172–186), slow cycling can be a form of reappropriation of time-space, contrasting its compression which is typical of our postmodern growth-oriented societies (Harvey 1989). When cycling, people become more aware of actual distances and often learn how to reduce their travel needs. Framing discourses about cycling as something naturally better aims not only to increase the number of people cycling and of people demanding safe cycling infrastructure. It aims to use cycling as one of the pivotal points from which a new ecosophy (Stibbe 2015) can be constructed, restoring the norms of sufficiency (see Popan 2019) that are necessary to make sure that economic and political processes 'will in fact constitute meaningful progress', as wished by Stiglitz (2004, 25).

As we observed in Chapter 5, presenting cycling as an individual choice brings the focus back on the person and moves away from road user labels and their negative effects. This is positive, but it is also necessary to bear in mind Mulderrig's caveats (2018) when she points out how the 5-a-day campaigns do not address the real societal causes behind obesity. Campaigns promoting velomobility must always beware of the fact that many people still choose a car because they feel constrained to do it by the road infrastructure, the commute they have to undertake and several other constraints brought about by the dominant contemporary lifestyles. Most cities are still planned around car use and making the traffic flow smoothly is too often considered an obvious public good, even when studies show it is the cause for increases in fatalities of people walking and cycling (see Baker 2019). It is thus necessary to strengthen the moral values arguing that life should be considered more important than increasing the sales of the latest SUV model, but at the same time frame it as a social issue rather than judging the individual choice of buying a new car or driving.

From a linguistic perspective, it is necessary to deconstruct the hegemony of automobility and the semiotic system of the roads, reinforcing the idea that space does not belong to cars but rather *exists* and *benefits* the living beings who *belong* and *relate to* it. Throughout the book we have provided some examples of reframing: 'road tax' could be replaced by 'pollution tax', as the official name for 'road tax'—'vehicle excise duty'—is not likely to ever become the everyday label for it and 'road tax' reinforces the idea that only those who pay it own the roads. We have seen the different labels employed to talk about lanes: referring more frequently to the lanes for people who drive as 'traffic lanes' can help to question the idea that cars are the norm and all other vehicles have special lanes dedicated to them. 'Special lanes' 'dedicated' certainly has a more positive prosody compared to 'segregated'. Moreover, we have shown the importance of reducing the prominence of the economic argument, for example by presenting the benefits as *socio*-economic. In Chapter 7, we highlighted the need to be aware of the detrimental effect that an excessive use of 'culture' can generate. And we have underlined how 'everyday cycling' is better to convey ideas of simplicity and avoids the unnecessary separation between 'utilitarian' and 'leisure' cycling.

The discursive challenge we face is striking a balance between our ingrained need to look for individual gain and the need to generate new, slower worldviews allowing humans to live with high well-being while keeping within the environmental limits. A possible way out is working on the reframing of well-being and what being well actually means, foregrounding the importance of contentment and its inextricable dependence on good health. While discourses promoting cycling strive to find effective ways of conveying the advantages of cycling, this discourse is paradoxically present—and effective—as a metaphor employed to advertise plenty of other goods and services that are totally unrelated to cycling. While this book has highlighted the downsides of marketisation, we must not throw the baby out with the dirty water and we can learn and use what copywriters have already done implicitly to promote the positivity of cycling by using the bicycle as a symbol of freedom, happiness and vitality.

Most campaigns inviting people to cycle more are envisaged from the perspective that Poli (2011, 166) defines 'moderate approach', aiming to change part of the mobility choices, but without questioning how much we feel we need to move. Reframing should be based on Poli's 'revolutionary approach' (168), which hinges on a reduction of our mobility needs. In order for this to be effective, issues of environmental protection and money saving need to be framed as positive side effects.

The most prominent aspect needs to become the well-being of people, grounding and radicating—two metaphors aptly related to the earth—the best choice in what feels most natural for our bodies. As argued by Latour (2018), we need to reframe the opposition between Left and Right and introduce the ecological threat as a central issue if we want to understand politics in the last fifty years. As we need to learn new ways to inhabit the Earth, Latour highlights the importance of new metaphorical thinking.

The promotion of cycling thus needs to be turned upside down, not a simple goal aiming to shift a number of people from one means of transport to another, but a radical revolution that brings to the surface the celebration of being alive and of well-being that, as humans, we seem to have forgotten, or at least subsided. In the time span in which this book was written, new movements like Fridays For Future and Extinction Rebellion were born. It is too early to say, but the number of people involved in them may mean that the widespread perception is actually changing. Cycling can thus be one of the pivotal points from which this new perspective can be framed, as it has the advantage of being something simple, something that many people enjoy without the need of identifying themselves as environmentalists. As explained by Lakoff (2010), it is necessary to create a public frame holding the economy and the environment together, rather than as opposites, showing that once the economy is brought back in its subordinate place in the system's hierarchy (see Meadows 2008) what is good for the economy is good for the environment and vice versa.

Language shapes society and is shaped by society: as we strive to find metaphors and linguistic features that can help us foster alternative worldviews, we need to rebuild our ways of conceiving society, our economic system and our unsustainable lifestyle. We need to ground what may appear as an overwhelming and unattainable goal into something familiar and already ingrained into our consciences. If, within our ecolinguistic perspective, reframing is one of our main goals, starting from the existing, positive frames is a reassuring starting point. Both English and Italian have an idiom testifying that cycling is second nature to our bodies, it is something we can't forget once we have learnt it, and it is often employed as an effective metaphor for simplicity and positivity. We can thus see it as a confirmation, as an invitation and as a reassuring positive note on which to close this investigation.

It's like riding a bike.

References

Baker P. C. (2019, October 3). Collision Course: Why Are Cars Killing More and More Pedestrians? *The Guardian*. https://www.theguardian.com/technology/2019/oct/03/collision-course-pedestrian-deaths-rising-driverless-cars. Accessed 23 November 2019.

Halliday, M. A. K. (2001). New Ways of Meaning: The Challenge to Applied Linguistics. In A. Fill & P. Mühlhäusler (Eds.), *The Ecolinguistics Reader: Language, Ecology and Environment* (pp. 175–202). London and New York: Continuum.

Harvey, D. (1989). *The Condition of Postmodernity: An Enquiry into the Origins of Cultural Change*. Oxford and Cambridge, MA: Wiley-Blackwell.

Lakoff, G. (2010). Why It Matters How We Frame the Environment. *Environmental Communication, 4*(1), 70–81.

Latour, B. (2018). *Down to Earth: Politics in the New Climatic Regime* (C. Porter, Trans.). Cambridge: Polity Press.

Meadows D. H. (2008). *Thinking in Systems: A Primer* (D. Wright, Ed.). White River Junction, VT: Chelsea Green Publishing.

Mulderrig, J. (2018). Multimodal Strategies of Emotional Governance: A Critical Analysis of 'Nudge' Tactics in Health Policy. *Critical Discourse Studies, 15*(1), 39–67.

Plehwe, D. (2016). Neoliberal Hegemony. In S. Springer, K. Birch, & J. MacLeavy (Eds.), *The Handbook of Neoliberalism* (61–72). London and New York: Routledge.

Poli, C. (2011). *Mobility and Environment: Humanists Versus Engineers in Urban Policy and Professional Education*. Heidelberg: Springer.

Popan, C. (2019). *Bicycle Utopias: Imagining Fast and Slow Cycling Futures*. London and New York: Routledge.

Purcell, M. (2016). Our New Arms. In S. Springer, K. Birch, & J. MacLeavy. (Eds.), *The Handbook of Neoliberalism* (613–620). London and New York: Routledge.

Sheller, M. (2018). *Mobility Justice: The Politics of Movement in an Age of Extremes*. London and New York: Verso Books.

Stibbe, A. (2015). *Ecolinguistics: Language, Ecology and Stories We Live By*. New York and London: Routledge.

Stiglitz, J. E. (2004). Evaluating Economic Change. *Daedalus Summer, 2004*, 18–25.

Glossary

Automobility Automobility is a complex amalgam of interlocking machines, social practices and ways of dwelling, not in a stationary home, but in a mobile, semi-privatized and hugely dangerous capsule.

Bikelash A perceived wave of 'angry' community opposition to new cycling infrastructure.

Concordance A display of every instance of a word in a corpus, showing context before and after.

Lemma A group of related wordforms, inflectional forms of the same base word. The verbs *cycling, cycle, cycles*, *cycled* are all part of the verb lemma *cycle*.

Nominalisation Verb processes that have been transformed into noun constructions—a transformation by which the people doing the action are removed, hiding both the agent and the affected.

Reification The action of regarding an abstract idea or concept as something concrete or real, having material existence.

Safety in numbers effect Describes the frequency of collisions in relation to the number of people walking and cycling. The relationship is not linear, and a rise in the number of people walking and cycling results in a higher decrease in the individual risk of being hit by a car.

Token A single wordform, all tokens that consist of exactly the same characters are examples of the same type. In the sentence 'the bicycle is in the garden' there are 6 tokens and 5 types.

M. C. Caimotto, *Discourses of Cycling, Road Users and Sustainability*, Postdisciplinary Studies in Discourse,
https://doi.org/10.1007/978-3-030-44026-8

Transitivity Who does what to whom and how. Through transitivity it is possible to background or omit responsibility, as agency and the responsibility for actions can be left implicit. Transitivity analysis is thus a powerful tool for analysing not only what is *in* texts but also what is *absent* from them.

Type Any single instance of an individual word in a text of corpus (compare 'token').

Index

M. C. Caimotto, *Discourses of Cycling, Road Users and Sustainability*, Postdisciplinary Studies in Discourse,
https://doi.org/10.1007/978-3-030-44026-8

The manufacturer's authorised representative in the EU is Springer Nature Customer Service Centre GmbH, Europaplatz 3, 69115 Heidelberg, Germany. If you have any concerns regarding our products, please contact ProductSafety@springernature.com

Printed and bound by CPI Group (UK) Ltd, Croydon, CR0 4YY
30/09/2026
02225609-0014